THE FUTURE OF NUCLEAR ENERGY

HEARING

BEFORE THE

SUBCOMMITTEE ON ENERGY,

COMMITTEE ON SCIENCE, SPACE, AND TECHNOLOGY

HOUSE OF REPRESENTATIVES

ONE HUNDRED THIRTEENTH CONGRESS

SECOND SESSION

December 11, 2014

Serial No. 113–99

Printed for the use of the Committee on Science, Space, and Technology

Available via the World Wide Web: http://science.house.gov

U.S. GOVERNMENT PUBLISHING OFFICE

WASHINGTON : 2015

92–332PDF

For sale by the Superintendent of Documents, U.S. Government Publishing Office
Internet: bookstore.gpo.gov   Phone: toll free (866) 512–1800; DC area (202) 512–1800
Fax: (202) 512–2104   Mail: Stop IDCC, Washington, DC 20402–0001

COMMITTEE ON SCIENCE, SPACE, AND TECHNOLOGY

HON. LAMAR S. SMITH, Texas, *Chair*

DANA ROHRABACHER, California
RALPH M. HALL, Texas
F. JAMES SENSENBRENNER, JR.,
   Wisconsin
FRANK D. LUCAS, Oklahoma
RANDY NEUGEBAUER, Texas
MICHAEL T. McCAUL, Texas
PAUL C. BROUN, Georgia
STEVEN M. PALAZZO, Mississippi
MO BROOKS, Alabama
RANDY HULTGREN, Illinois
LARRY BUCSHON, Indiana
STEVE STOCKMAN, Texas
BILL POSEY, Florida
CYNTHIA LUMMIS, Wyoming
DAVID SCHWEIKERT, Arizona
THOMAS MASSIE, Kentucky
KEVIN CRAMER, North Dakota
JIM BRIDENSTINE, Oklahoma
RANDY WEBER, Texas
CHRIS COLLINS, New York
BILL JOHNSON, Ohio

EDDIE BERNICE JOHNSON, Texas
ZOE LOFGREN, California
DANIEL LIPINSKI, Illinois
DONNA F. EDWARDS, Maryland
FREDERICA S. WILSON, Florida
SUZANNE BONAMICI, Oregon
ERIC SWALWELL, California
DAN MAFFEI, New York
ALAN GRAYSON, Florida
JOSEPH KENNEDY III, Massachusetts
SCOTT PETERS, California
DEREK KILMER, Washington
AMI BERA, California
ELIZABETH ESTY, Connecticut
MARC VEASEY, Texas
JULIA BROWNLEY, California
ROBIN KELLY, Illinois
KATHERINE CLARK, Massachusetts

———

SUBCOMMITTEE ON ENERGY

HON. CYNTHIA LUMMIS, Wyoming, *Chair*

RALPH M. HALL, Texas
FRANK D. LUCAS, Oklahoma
RANDY NEUGEBAUER, Texas
MICHAEL T. McCAUL, Texas
RANDY HULTGREN, Illinois
THOMAS MASSIE, Kentucky
KEVIN CRAMER, North Dakota
RANDY WEBER, Texas
LAMAR S. SMITH, Texas

ERIC SWALWELL, California
ALAN GRAYSON, Florida
JOSEPH KENNEDY III, Massachusetts
MARC VEASEY, Texas
ZOE LOFGREN, California
DANIEL LIPINSKI, Illinois
KATHERINE CLARK, Massachusetts
EDDIE BERNICE JOHNSON, Texas

CONTENTS

December 11, 2014

Appendix I: Answers to Post-Hearing Questions

THE FUTURE OF NUCLEAR ENERGY

THURSDAY, DECEMBER 11, 2014

House of Representatives,
Subcommittee on Energy
Committee on Science, Space, and Technology,
Washington, D.C.

The Subcommittee met, pursuant to call, at 10:08 a.m., in Room 2318 of the Rayburn House Office Building, Hon. Cynthia Lummis [Chairwoman of the Subcommittee] presiding.

LAMAR S. SMITH, Texas
CHAIRMAN

EDDIE BERNICE JOHNSON, Texas
RANKING MEMBER

Congress of the United States

House of Representatives

COMMITTEE ON SCIENCE, SPACE, AND TECHNOLOGY

2321 RAYBURN HOUSE OFFICE BUILDING

WASHINGTON, DC 20515-6301

(202) 225-6371
www.science.house.gov

Subcommittee on Energy

The Future of Nuclear Energy

Thursday, December 11, 2014
10:00 a.m. – 12:00 p.m.
2318 Rayburn House Office Building

Witnesses

Panel I

The Honorable Peter Lyons, Assistant Secretary, Office of Nuclear Energy, U.S. Department of Energy

Panel II

Dr. Ashley Finan, Senior Project Manager, Energy Innovation Project, Clean Air Task Force

Mr. Mike McGough, Chief Commercial Officer, NuScale Power

Dr. Leslie Dewan, Co-founder and Chief Executive Officer, Transatomic Power

Mr. Daniel Lipman, Executive Director, Policy Development, Nuclear Energy Institute

U.S. HOUSE OF REPRESENTATIVES
COMMITTEE ON SCIENCE, SPACE, AND TECHNOLOGY
SUBCOMMITTEE ON ENERGY

HEARING CHARTER

The Future of Nuclear Energy

Thursday, December 11, 2014
10:00 a.m. – 12:00 p.m.
2318 Rayburn House Office Building

Purpose

The Energy Subcommittee of the House Committee on Science, Space, and Technology will hold a hearing titled *The Future of Nuclear Energy*, at 10:00 a.m. on Thursday, December 11th in room 2318 of the Rayburn House Office Building. This hearing will discuss the next generation of reactor designs, the DOE's support through its Office of Nuclear Energy (NE), and challenges for private investment in new nuclear energy technology.

Witnesses

Panel I
- **The Honorable Peter Lyons**, Assistant Secretary for Nuclear Energy, U. S. Department of Energy

Panel II
- **Dr. Ashley Finan**, Senior Project Manager, Energy Innovation Project, Clean Air Task Force
- **Mr. Mike McGough**, Chief Commercial Officer, NuScale Power
- **Dr. Leslie Dewan**, Co-founder and Chief Executive Officer, Transatomic Power
- **Mr. Daniel Lipman**, Executive Director, Policy Development, Nuclear Energy Institute

Background

NE's mission is to advance nuclear power as a resource capable of meeting the United States' energy supply, environmental, and national security needs. NE's total funding for civilian nuclear energy R&D totaled approximately $489 million in fiscal year 2014. NE supports this effort through its R&D programs, including Small Modular Reactor (SMR) Technical Support, Reactor Concepts, Fuel Cycle R&D, and Nuclear Energy Enabling Technologies (NEETs).

4

The purpose of NE's Reactor Concepts program is to develop new reactor designs, including high-temperature gas-cooled reactors (HTGRs), liquid metal-cooled reactors, and liquid salt-cooled reactors. These advanced designs present numerous advantages over the current fleet, such as yielding higher temperatures for industrial applications, burning nuclear waste as fuel, minimizing the volume of waste products, and overall increased efficiency and safety. This program also supports, among other things, research to support life extensions of existing reactors (large light water reactors). DOE requested approximately $100 million for this program in fiscal year 2015.

The SMR Technical Support program supports first-of-a-kind costs for design certification and licensing activities through the Nuclear Regulatory Commission (NRC) on a minimum 50% cost-shared basis with industry partners.[1] DOE requested approximately $97 million for this program in fiscal year 2015.

The Fuel Cycle R&D program supports, among other things, research to reduce waste, enhance safety, and limit proliferation risk from the nuclear fuel cycle. Pursuant to the DOE's *Strategy for the Management and Disposal of Used Nuclear Fuel and High Level Radioactive Waste*,[2] the Fuel Cycle R&D program funds R&D related to storage, transportation, and disposal of used nuclear fuel, $24 million of which is sourced from the Nuclear Waste Fund.[3] For fiscal year 2015, DOE requested approximately $189 million for the Fuel Cycle R&D program.

The Nuclear Energy Enabling Technologies (NEETs) program funds R&D in crosscutting technology areas, including materials, sensors and instrumentation, and advanced manufacturing, as well as modeling and simulation of reactor systems. DOE requested approximately $78 million for this program in fiscal year 2015.

Currently, the United States generates approximately 20 percent of its electricity from nuclear reactors that use water, also known as "light water"[4] in the nuclear context, to both cool the reactor and slow down neutrons to fission atoms of uranium fuel. Dating back to the 1950s, the United States began development and construction of numerous advanced reactor designs for research purposes. These non-light water reactor designs can reach higher levels of thermal efficiency, some of which can use nuclear waste as fuel, including fast reactors, high temperature gas-cooled reactors, and liquid salt-cooled reactors. Transatomic, a company located in Cambridge, Massachusetts, intends to develop a molten-salt reactor design that will "turn nuclear waste into a safe, clean, and scalable source of electricity."[5]

---

[1] See generally, DOE website: http://www.energy.gov/ne/nuclear-reactor-technologies/small-modular-nuclear-reactors; See also DOE cost sharing for R&D contracting here: http://energy.gov/sites/prod/files/35.2_Cost_Sharing_in_Research_and_Development_Contracts_0.pdf

[2] http://energy.gov/sites/prod/files/Strategy%20for%20the%20Management%20and%20Disposal%20of%20Used%20Nuclear%20Fuel%20and%20High%20Level%20Radioactive%20Waste.pdf

[3] In accordance with the Nuclear Waste Policy Act of 1982, as amended, the Nuclear Waste Fund is derived from fees imposed on nuclear power utilities, which carries a current balance of approximately $30 billion.

[4] Light Water refers to water composed of the more common isotope of hydrogen with one proton and zero neutrons per atom. Heavy water refers to water composed of heavy hydrogen which has one proton and one neutron per atom.

[5] http://transatomicpower.com/

In the late 1950s, the U.S. Navy began its nuclear program to fuel submarines and eventually aircraft carriers by way of small reactors. In general, small modular reactors would allow for lower initial investment compared to larger reactors, scalability, and siting flexibility. NuScale Power, a company based in Corvallis, Oregon, is developing an SMR with passive safety features that will cool itself down in an accident scenario without the need for any electricity or mechanized systems.[6]

Additional Reading:

- U.S. Government Accountability Office, *"Advanced Reactor Research: DOE Supports Multiple Technologies, but Actions Needed to Ensure a Prototype Is Built"* (June, 2014), available here: http://www.gao.gov/products/GAO-14-545

---

[6] http://www.nuscalepower.com/

Chairwoman LUMMIS. The hearing of the Subcommittee on Energy will come to order.

Good morning and welcome to today's joint hearing titled ''The Future of Nuclear Energy.'' In front of each Member are packets containing the written testimony, biographies, and truth-in-testimony disclosures for today's witnesses.

And before I give an opening statement, I would like to say that it has been my pleasure for the past two years to serve on a Committee that is Chaired by one of the most distinguished Members of Congress who has been the type of Chairman that every Member of Congress hopes that they will have the opportunity to serve with and under. Lamar Smith is a gentleman's gentleman and has been one of the most wonderful people that a Subcommittee Chairman could have the opportunity to tutor under.

This Committee has conducted I believe 87 hearings during the course of the last two years under Chairman Smith's leadership and we have done it in a manner that has been respectful, that has sought information, allowed us to be better advocates and more knowledgeable Members of Congress.

So I would like to start before I give my opening statement to take this opportunity to thank Chairman Smith for his leadership, his mentorship, and for his many years in Congress from which all of us who have worked with him and under him have benefited.

Mr. WEBER. Will the gentlelady yield?

Chairwoman LUMMIS. I will yield.

Mr. WEBER. I have got to get ahead of our Chairman here. I just want to say ditto.

Chairman SMITH. You all are nice. Would the Chair yield just for a minute?

Chairwoman LUMMIS. The Chair will yield.

Chairman SMITH. I just want to thank Cynthia Lummis for being an outstanding Subcommittee Chairman. We will miss her but always support her in her other committee assignments. And she has not only been an outstanding Chair, she has been an outstanding Member of Congress. She is knowledgeable, she is conscientious, she is thoughtful, she is diplomatic, but she is also strong, and those qualities have made Cynthia Lummis one of the outstanding Members of Congress that we have today.

So, Cynthia, thank you very much for all you have done and thanks for spreading good rumors about me.

Chairwoman LUMMIS. Thank you. Thank you, Mr. Chairman.

The Members of this Committee which will participate under your leadership during the next Congress are fortunate indeed to have you as their leader and mentor.

Well, good morning. And I would like to welcome our witnesses for today's hearing. Today, we will look at the track record and road forward for research and development within DOE's Office of Nuclear Energy. We will also look at the progress of nuclear energy technology in the United States and the regulatory environment for licensing new reactors.

Nuclear power currently accounts for approximately 19 percent of the United States' electricity generation and 60 percent of our emission-free electricity. And my home State of Wyoming is the nation's largest producer of uranium. Nuclear energy is reliable, resilient, and has safely powered America for decades.

But the fundamental questions about the future of this technology need to be answered: When will we see the commercialization of small modular nuclear reactors that can be deployed at offgrid locations, something of particular interest I might say from me coming from the most sparsely populated state in the United States. When will we see deployment of advanced reactors that can reach much higher levels of thermal efficiency, recycle nuclear waste, and serve as hybrid energy systems? And what are the regulatory and market barriers slowing down progress of these technologies in the United States?

As many of us know, the time frame for bringing a nuclear reactor online is unforgivably long and so we must work together to make sure that we can make it more time-sensitive.

Nuclear energy was born in the United States. We have the best scientists and engineers in the world. We are fortunate to have some of them here today. Yet we are not seeing the pace of commercial technology advancement that we would expect. At the same time, other countries, including China, are surging ahead.

We have to ask ourselves: Is the United States going to remain a global leader in nuclear technology? These are the issues that we want to discuss today. I look forward to further discussion and again, I thank the witnesses for participating in today's hearing.

[The prepared statement of Mrs. Lummis follows:]

PREPARED STATEMENT OF SUBCOMMITTEE ON ENERGY
CHAIRWOMAN CYNTHIA LUMMIS

Good morning. I would like to welcome our witnesses to today's hearing. Today, we will look at the track record and road forward for research and development within DOE's Office of Nuclear Energy. We will also look at the progress of nuclear energy technology in the United States and the regulatory environment for licensing new reactors.

Nuclear power currently accounts for approximately 19% the United States' electricity generation and 60% of our emission-free electricity. And, my home state of Wyoming is the nation's largest producer of uranium. Nuclear energy is a reliable, resilient, and has safely powered America for decades.

But, fundamental questions about the future of this technology need to be answered:

- When will we see the commercialization of small modular reactors that can be deployed at offgrid locations?
- When will we see deployment of advanced reactors that can reach much higher levels of thermal efficiency, recycle nuclear waste, and serve as hybrid energy systems?
- What are the regulatory and market barriers slowing down progress of these technologies in the United States?

Nuclear energy was born in the United States. We have the best scientists and engineers in the world. Yet, we are not seeing the pace of commercial technology advancement that we would expect. At the same time, other countries including China are surging ahead.

We have to ask ourselves: is the United States going to remain a global leader in nuclear technology? These are the issues we intend to discuss today. I look forward to further discussion and again, I thank the witnesses for participating in today's hearing.

Chairwoman LUMMIS. The Chair now recognizes our Ranking Member Mr. Swalwell for his opening statement.

Mr. SWALWELL. Thank you, Chairman Lummis.

And first, I would also like to express my good wishes for you going forward. I have enjoyed working with you. You were very kind early on when we both were selected to lead our respective sides on this Subcommittee. And we met and we have talked about what our mutual interests are, I think especially in an all-of-the-above energy approach I do think you have led this Committee every time with an open mind, with dignity, and it is something that I will miss. But I know that there are many great things ahead for what you will do and the work you will accomplish in the Congress.

So thank you for just being so gracious. Even during contentious hearings, you never ceased to allow both sides to be heard and you were always open, I think, to whatever ideas were out there that could move our country forward and I really do appreciate that.

Chairwoman LUMMIS. Thank you. Will the gentleman yield?

Mr. SWALWELL. Yes.

Chairwoman LUMMIS. I, too, want to acknowledge what a proper and important and dignified and lovely working relationship that I have had with the Ranking Member Mr. Swalwell. It has been a breath of fresh air. And we are good partners on this Committee and I have very much enjoyed our working relationship and my very best to you.

Mr. SWALWELL. And, Chair, I will never forget you asked me one time during a meeting—I told you I believe when it comes to energy if we can make it safe, we should make it happen, and you looked me in the eye and you said do you really believe that? And I said yes, I do. And you said I am going to put the screws to you on that and hold you to that and you have, and I appreciate that. And I hope we can find more ways where we both believe we can make it safe, we can make it happen.

And speaking of making it safe and making it happen, with respect to today's hearing, for decades, the federal government has provided critical support for energy R&D. And from solar and wind to natural gas recovery, many of the technologies that are helping us transition to a clean energy economy and creating entire new industries wouldn't be nearly as far along as they are today, or would not exist without the benefit of the partnerships between the federal government and public and semi-public partnerships and entities.

I look forward to learning more today about nuclear energy, particularly from our witness from NuScale power, who we will be hearing from today, who has been working with Sandia National Laboratory, which is located in Livermore, California, in the 15th Congressional District, which I have the privilege to represent.

This morning we are here to discuss the federal role in the development and deployment of the next generation of nuclear power plants and how this support may be better structured going forward. I am eager to learn about the costs and benefits of these new technologies over the course of the hearing, including ways we can improve the safety of new reactors to minimize the chance of another catastrophic event along the lines of the disaster that occurred at the Fukushima plant just a few years ago.

I have stated a number of times that I just referenced that I believe and support an all-of-the-above "if we can make it safe, we should make it happen'' approach to clean energy, and achieving a safer, more cost-effective and environmentally friendly way to utilize nuclear energy, and how that can play an important role in this mix. We just need to make sure that we are making the smartest investments we can with our limited, challenged resources and that they are in the best interest of the American people.

Again, I want to thank the witnesses, particularly Dr. Lyons, today for being willing to provide their insights. I look forward to working with my colleagues on the other side of the aisle and with all of the stakeholders in this critical, critical area moving forward.

Again, thank you, Chairman Lummis, and I yield back.

[The prepared statement of Mr. Swalwell follows:]

PREPARED STATEMENT OF SUBCOMMITTEE ON ENERGY
RANKING MINORITY MEMBER ERIC SWALWELL

Thank you Chairman Lummis for holding this hearing, and I also want to thank this excellent panel of witnesses for their testimony and for being here today.

For decades, the federal government has provided critical support for energy R&D. From solar and wind energy to natural gas recovery, many of the technologies that are helping us transition to a clean energy economy and creating entire new industries wouldn't be nearly as far along as they are today, or would not exist at all, without the benefit of federal support and public-private partnerships. The same certainly holds true for nuclear energy and in fact, NuScale Power, who we'll be hearing from today, has been working with Sandia National Laboratories.

This morning we are here to discuss the federal role in the development and deployment of the next generation of nuclear power plants, and how this support may be better structured going forward. I'm eager to learn more about the costs and benefits of these new technologies over the course of the hearing—including ways we can improve the safety of new reactors to minimize the chance of another catastrophic event along the lines of the disaster that occurred at Fukushima just a few years ago.

I have stated numerous times that I support an ''all of the above'' approach toward a clean energy economy and achieving safer, more cost-effective, and environmentally friendly ways to utilize nuclear energy can play an important role in this mix. We just need to make sure that we are making the smartest investments we can with our limited resources, and that they are in the best interests of the American people. I want to thank the witnesses again for being willing to provide their insights today, and I look forward to working with my colleagues on the other side of the aisle and with all of the stakeholders in this critical area moving forward.

Thank you again, Chairman Lummis, and I yield back.

Chairwoman LUMMIS. Thank you, Mr. Swalwell.

I now recognize the Chairman of the full Committee for a statement.

Chairman SMITH. Okay. Thank you, Madam Chair.

Today's hearing will examine both current and future challenges and opportunities that face nuclear power.

Nuclear power is a proven source of emission-free electricity that has been generated safely in the United States for over half a century. However, our ability to move from R&D to market deployment has been hampered by government red tape and partisan politics. We are just now seeing the first reactors under construction in more than 30 years. This hiatus has diminished our supply chain and ability to build new reactors. In fact, the United States no longer has the capability to manufacture large reactor pressure vessels.

Today, we will hear from NuScale, a company that is the closest to navigating the Nuclear Regulatory Commission's licensing process to build and deploy the first small modular reactors in the United States, a subject that our colleague Dana Rohrabacher has long been interested in.

We will also hear from Transatomic, a company recently formed by two graduate students from MIT that could revolutionize the energy sector. Transatomic's technology would recycle spent nuclear fuel, achieve higher levels of efficiency than existing designs, and yield minimum radioactive byproducts.

The United States has not lived up to its potential when it comes to nuclear energy. The regulatory process is cumbersome and lacks the certainty needed for sustained investment in new nuclear energy technology. I am hopeful that this hearing can serve as a forum for how to enable nuclear power to meet more of our energy needs.

Thank you, Madam Chair. I yield back.

[The prepared statement of Mr. Smith follows:]

PREPARED STATEMENT OF FULL COMMITTEE CHAIRMAN LAMAR S. SMITH

Today's hearing will examine both current and future challenges and opportunities that face nuclear power.

We will first hear from the Department of Energy on its research and development (R&D) strategy to ensure the United States' nuclear energy industry remains competitive. Our second panel will discuss the challenges that developers face in today's regulatory environment. Nuclear power is a proven source of emission-free electricity that has been generated safely in the United States for over half a century.

However, our ability to move from R&D to market deployment has been hampered by government red tape and partisan politics. We are just now seeing the first reactors under construction in more than 30 years.

This hiatus has diminished our supply chain and ability to build new reactors. In fact, the United States no longer has the capability to manufacture large reactor pressure vessels.

Today, we will hear from NuScale, a company that is the closest to navigating the Nuclear Regulatory Commission's licensing process to build and deploy the first small modular reactors in the United States.

We will also hear from Transatomic, a company recently formed by two graduate students from MIT that could revolutionize the energy sector.

Tranastomic's technology would recycle spent nuclear fuel, achieve higher levels of efficiency than existing designs, and yield minimal radioactive byproducts. The U.S. has not lived up to its potential when it comes to nuclear energy. The regulatory process is cumbersome and lacks the certainty needed for sustained investment in new nuclear energy technology.

I am hopeful that this hearing can serve as a forum for how to enable nuclear power to meet more of our energy needs.

Chairwoman LUMMIS. I thank the Chairman.

If there are Members who wish to submit additional opening statements, your statement will be added to the record at this point.

Chairwoman LUMMIS. It is now time to introduce our first witness panel. Our first witness today is thw Honorable Peter Lyons, Assistant Secretary for the Office of Nuclear Energy at the Department of Energy. Dr. Lyons previously served as Principal Deputy Assistant Secretary for the Office of Nuclear Energy. Prior to joining DOE, Dr. Lyons was the Commissioner of the Nuclear Regulatory Commission focusing on safety and operating reactors.

As our witnesses should know, spoken testimony is limited to five minutes each after which Members of the Committee have five minutes each to ask questions. Your written testimony will be included in the record of the hearing.

So without further ado I now recognize our witness Dr. Lyons.

TESTIMONY OF THE HONORABLE PETER LYONS, ASSISTANT SECRETARY, OFFICE OF NUCLEAR ENERGY, U.S. DEPARTMENT OF ENERGY

Mr. LYONS. Thank you very much.

Chairman Lummis, Ranking Member Swalwell, and Members of the Committee, thank you for your invitation to testify at the Committee's hearing today on the future of nuclear energy.

Nuclear energy continues to play a vital role in President Obama's all-of-the-above energy strategy for a sustainable clean energy future. Nuclear energy has provided nearly 20 percent of our electrical generation over the past two decades and now produces over 60 percent of our zero carbon electricity.

In order for nuclear energy to continue this role, the Office of Nuclear Energy, or NE, focuses on programs to improve the reliability, performance, and operating lifetime of current reactors, support the deployment of affordable advanced reactors, develop a sustainable nuclear fuel cycle, maintain key infrastructure, and manage international collaborations.

The current light water reactor, or LWR fleet, is challenged by economic conditions that contributed to the early closure of four reactors in 2013 in addition to the imminent retirement of the Vermont Yankee plant. The shutdown of these power plants is a significant loss of low carbon electricity. Nevertheless, we remain optimistic with the current construction of five nuclear reactors, four of which are the Westinghouse AP1000, a new generation of passively safe reactors. Two of these plants received over $6 billion in loan guarantees, and for future assistance, the Department recently released a $13 billion loan guarantee solicitation for advanced nuclear energy projects.

In conjunction with industry and, more appropriate, the NRC, the LWR Sustainability Program supports the current fleet for possible license renewals beyond 60 years, and this program also addresses the lessons learned from the Fukushima Daiichi accident.

A high priority of the Department is to accelerate the commercialization and deployment of small modular reactors or SMRs with our Cost-Shared Licensing Technical Support Program. SMRs can promote American competitiveness, create domestic manufacturing jobs, and help reduce CO_2 emissions. The two small modular LWRs supported by the Department feature extremely impressive passive safety.

Future reactor systems may employ advanced designs to improve performance beyond what is currently available. Coolants other than light water may enable reactors to operate at higher temperatures with improved efficiencies and economics, as well as optimize their waste forms. The Department has supported industrial R&D on these advanced reactor designs through cost-shared agreements, as well as supported R&D at national labs and universities. In addition, we also continue to leverage international experience through the Generation IV International Forum.

Progress towards a consent-based solution to managing the nation's nuclear waste and used fuel remains a challenge that must be addressed. In January 2013 the Administration released its strategy for this task. And pursuant to that strategy, my office is undertaking activities within its existing authority to plan for the eventual transportation, storage, and disposal of used nuclear fuel, as well as R&D on related topics.

By way of conclusion, any programs encompass all aspects of nuclear power including support for the nation's 100 operating LWRs which remain a vital national resource of safe, clean energy but new plants are also needed. Past programs like the cost-shared NP 2010 program provided two certify designs for passively safe, large LWRs, and in an analogous way, our current licensing technical support program strives to provide design certification for two SMRs. If we are successful with that program, the nation will have two complementary approaches to new plant construction well matched to the wide range of our domestic needs, as well as addressing international markets.

In planning for future advanced reactors it is appropriate to remember the words of Hyman Rickover when he discussed the differences between paper and real reactors. He noted the challenges of bringing a new reactor design online are substantial and are hard to fully anticipate as the project is planned. His words are not a reason to forgo development of advanced reactors but they should remind us of the challenges inherent in such endeavors even though several of the advanced concepts have some operational history.

In the United States we have comprehensive knowledge of LWRs. We can design and regulate them with highest confidence for safe operations. Today's advanced concepts will be deployed only if they are based on the same confidence that we have today for LWRs. Research today should focus on providing that level of confidence for these new concepts for tomorrow.

To use advanced reactors in the future we need to maintain a strong domestic nuclear energy industry, including utilities, with operational experience on nuclear systems. In the near term the latest generation of LWRs and the promising new SMRs must serve as an essential bridge between the reactors of today and the future potential for new reactor designs. And without that bridge any path towards non-light water reactors will be challenging.

Thank you and I look forward to your questions.

[The prepared statement of Mr. Lyons follows:]

Statement of Dr. Peter Lyons, Assistant Secretary for Nuclear Energy
U.S. Department of Energy
Before the
Committee on Science, Space and Technology
Subcommittee on Energy
U.S. House of Representatives
December 11, 2014

Chairman Lummis, Ranking Member Swalwell, and members of the Committee, thank you for your invitation to testify at today's hearing on "The Future of Nuclear Energy." Nuclear energy continues to play a vital role in President Obama's "all-of-the-above" energy strategy for a sustainable, clean energy future. Nuclear energy has provided nearly 20 percent of electrical generation in the United States over the past two decades and now produces over 60 percent of America's zero-carbon electricity. As a deployable power source with a high capacity factor, nuclear power fits the needs for baseload power – and does it with today's technologies. A prerequisite for nuclear power continuing as a vital part of the nation's clean energy portfolio is public confidence in the safety of nuclear plants and commercial confidence that the plants can be operated safely, reliably, and economically. In order to ensure that nuclear energy continues to provide affordable, carbon-free power, the Office of Nuclear Energy (NE) focuses its programs to: improve the reliability and performance, sustain the safety and security, and extend the life of current reactors by developing advanced technological solutions; meet the Nation's energy security and climate change goals by developing technologies to support the deployment of affordable advanced reactors; improve energy generation, waste management, safety, and nonproliferation attributes by developing sustainable nuclear fuel cycles; maintain key infrastructure to support cutting edge research on nuclear technologies; and advance U.S. civil nuclear energy priorities and objectives through international collaboration.

The Current Fleet

NE works in conjunction with industry and, where appropriate, with the Nuclear Regulatory Commission (NRC) to support and conduct the long-term research needed to inform major component refurbishment and replacement strategies for the current fleet. These areas include performance, cyber security, and safety; long-term operations through plant license extensions; and age-related regulatory oversight decisions.

One of NE's key programs, the Light Water Reactor Sustainability (LWRS) program, addresses challenges facing the continued safe and economic operation of the current fleet. The LWRS program focuses research on material aging issues. Its findings will inform license renewal applications for operation beyond 60 years, which may be submitted by industry starting in the 2016 to 2018 time period. Extending the operating lifetimes of current plants beyond 60 years and, where possible, making further improvements in their productivity will generate near-term benefits. Activities in this area have been expanded to address lessons learned from the Fukushima Daiichi accident, particularly in understanding and managing Severe Accident (SA) events. These include evaluation of instrumentation to better monitor and manage SAs, computer analysis of SA progression, and preparation and planning efforts in support of eventual examination of the damaged reactors. The LWRS program has partnered with industry to closely coordinate research needs and share costs. The program also coordinates with the NRC to improve the utility of research results.

Technical questions are not the only issues facing the current reactor fleet. Adverse economic conditions contributed to the early closure of four reactors in 2013 and are a factor in the imminent retirement of the Vermont Yankee plant. Complex market factors, falling alternative generations costs, and lower electricity demand forecasts have made operating nuclear power plants uneconomical in some parts of the country. Several more reactors may be at risk of early closure due to these economic forces and the increasing costs of operation. The shutdown of these power plants is a significant blow to zero-carbon electricity generation as well as a considerable loss of baseload electricity supply and energy diversity. America's nuclear power fleet is a national asset on many fronts, and our programs work to ensure nuclear remains a key player in America's clean energy future.

Licensing and Construction of Nuclear Reactors in the United States

Although the United States has experienced a reduction in nuclear capacity with these plant closings, we see cause for optimism with the current construction of five nuclear reactors. Tennessee Valley Authority's (TVA) Watts Bar Unit 2 in Tennessee is about 90 percent complete and, with a targeted in-service date of December 2015, will be the first U.S. reactor to be completed this century. Construction of the first new nuclear plants in this country in more than 30 years continues for two new units at VC Summer in South Carolina and two new units at Plant Vogtle in Georgia. Both projects are deploying the NRC-certified, Generation III+ Westinghouse AP1000, a new generation of passively safe reactors. Earlier this year, the Department of Energy's Loan Programs Office announced that two of the owners of Plant Vogtle received a $6.5 billion loan guarantee to support construction of the Vogtle facility. Together, these newly constructed units will provide enough reliable, zero-emission, baseload electricity to power three million homes in the Southeastern United States, with current estimates for completion projected in the 2017 to 2019 timeframe.

Further headway has been made with the recent certification by the NRC of the GE-Hitachi Nuclear Energy's Economic Simplified Boiling Water Reactor (ESBWR). The ESBWR is a 1,600 megawatt reactor, which includes passive safety features that would cool the reactor after an accident without the need for human intervention. The design is currently being considered for deployment in Virginia and Michigan.

If nuclear energy is to continue to be a strong component of the Nation's energy portfolio, barriers to the further deployment of new nuclear plants must be overcome. Impediments to new plant deployment, even for those designs based on familiar Light Water Reactor (LWR) technology, include the substantial capital cost of new plants and the uncertainties in the time required to license and construct those plants.

A high priority of the Department has been to accelerate the timelines for the commercialization and deployment of small modular reactor (SMR) technologies through the SMR Licensing Technical Support (LTS) program. The SMR LTS program is a six-year, $452 million initiative focused on first-of-a-kind engineering support for design certification and licensing activities for SMR designs through cost-shared arrangements with industry partners to promote accelerated commercialization of the nascent technology. SMRs have the potential to achieve lower upfront capital cost, modular power additions, and simpler, predictable and faster construction than other designs. The Department believes strongly that SMRs can promote American competitiveness, create manufacturing jobs here at home, and reduce CO_2 emissions through clean, safe, and reliable nuclear power.

These new SMRs, as well as the AP1000 and ESBWR reactors, are designed with passive safety features to minimize any requirement for prompt operator action and to prevent auxiliary system failures from contributing to future accidents. These attributes further enhance the safety of nuclear power plants. Overall, the SMR Program supports the licensing of innovative designs that improve safety, operations and economics. We expect these SMRs to have lower core damage frequencies, longer post-accident coping periods, enhanced resistance to natural phenomena, and potentially smaller emergency preparedness zones than currently licensed reactors.

The Department has entered into two cost-shared agreements with industry. In November 2012, DOE announced the selection of the mPower America team, consisting of Babcock & Wilcox (B&W), Bechtel International and the Tennessee Valley Authority, for cost-shared investment to support the design development, certification, and licensing activities of B&W's mPower reactor to be sited at TVA's Clinch River site in Tennessee by 2022. In May 2014, the Department of Energy signed a cooperative agreement with NuScale Power, providing $217M in DOE funds to support design development and NRC design certification with deployment scheduled for the 2025 timeframe.

The Department also recognizes the challenge of constructing large capital nuclear projects. The Department issued a loan guarantee for the Vogtle project, which represented an important step in deploying advanced nuclear technology. However, other innovative nuclear projects may be unable to obtain full commercial financing due to the perceived risks associated with technology that has never been deployed at commercial scale in the United States. To that end, in September 2014, the Department's Loan Programs Office released a draft $12.6 billion loan guarantee solicitation for advanced nuclear energy projects. The loan guarantees from this draft solicitation would support advanced nuclear energy technologies including advanced nuclear reactors, SMRs, uprates and upgrades at existing facilities, and front-end nuclear projects. The Department accepted public comments for 30-days following publication of the draft solicitation; those comments are currently under review.

Research and Development for Advanced Reactor Technologies

Future-generation reactor systems may employ advanced technologies and designs to improve performance beyond what is currently attainable. More advanced reactor designs with coolants other than light water, often referred to as Generation IV designs, may enable reactors to operate at higher temperatures and with increased efficiencies resulting in improved economics. These designs may also provide expanded fuel cycle options that can inform future policy decisions. Generation IV reactor designs are being developed in many countries, and research, development, and demonstration continues in the United States in support of a number of these concepts. Continued strong research and development in this area is essential for the long-term prospects of nuclear energy.

The Department's advanced reactor program performs research to develop technologies and subsystems that are critical for advanced concepts that could dramatically improve nuclear power performance through the achievement of goals on sustainability, economics, safety, and proliferation resistance. Advanced reactor technologies considered in this program reside at different maturity levels with research and development efforts mainly focused on three advanced concepts: liquid metal-cooled fast reactors, including sodium-cooled fast reactors (SFRs); fluoride salt-cooled high-temperature reactors (FHRs); and high-temperature gas-cooled reactors (HTGR). In addition, research and development addresses qualification of tristructural-isotropic (TRISO) coated particle fuel and graphite used in both FHRs and HTGRs.

The Department of Energy has issued several awards over the last two years to support industry's research and development activities through cost-share agreements totaling $16.5 million in government funding. These projects will help address significant technical challenges to the design, construction, and operation of next generation nuclear reactors, based upon the research and development needs identified by industry designers and technical experts. In many cases, new technologies will be needed to enable advanced reactor designs.

To further support the development of advanced reactor technologies, the Department has undertaken a joint initiative with the NRC to support development of General Design Criteria (GDC) for Nuclear Power Plants for advanced reactor designs. The current GDC were initially developed primarily for light water reactors. .

NE also continues to leverage international experience through the Generation IV International Forum (GIF) where the United States collaborates on important research for HTGRs and SFRs. Over the past ten years, GIF has served as a unique multilateral mechanism for coordinating research and development on advanced reactors, resulting in the completion of hundreds of research deliverables and milestones in such areas as irradiation and fuel development, materials, chemistry, safety, and reactor operations.

The Department has also begun to study how to optimize nuclear energy with variable renewable energy sources through collaboration between the Offices of Nuclear Energy and Energy Efficiency and Renewable Energy. These studies will not only examine integration of current light water reactor technology, but also advanced reactor technologies that have the potential to provide high temperature process heat in addition to higher efficiency electricity. This can theoretically be accomplished by installing a combination of additional nuclear and renewable energy production technologies, improving or developing new energy storage capacity, and using excess capacity in the electric sector to provide clean thermal and/or electrical energy to the manufacturing and fuels industries. For the past ten years, NE has conducted research on supercritical carbon dioxide (sCO_2) Brayton cycles for use - with advanced reactor concepts. Recent efforts to accelerate the commercialization of this transformational energy conversion technology have been proposed in the Supercritical Transformational Electric Power

Generation initiative. This proposed initiative, which would build upon program research and development efforts by the Offices of Nuclear Energy, Fossil Energy, and Energy Efficiency and Renewable Energy, intends to build a 10 megawatt demonstration facility under a 50/50 cost-share with industry. This technology is suitable for a variety of applications including concentrating solar power, geothermal, advanced nuclear technologies (HTGR and SFR), and fossil fuels with efficiencies higher than traditional steam cycles using Rankine cycle technology currently used in those applications.

Sustainable Fuel Cycle

Finding a long-term, consent-based solution to managing the nation's nuclear waste and used nuclear fuel (UNF) is a long standing challenge. Such a solution, however, is necessary to assure the future viability of this important carbon-free energy supply and further strengthen America's standing as a global leader on issues of nuclear safety and nonproliferation.

In January 2013, the Administration released its *Strategy for the Management and Disposal of Used Nuclear Fuel and High-Level Radioactive Waste*, which lays out plans to implement a long-term program that includes development of a pilot interim storage facility, a larger consolidated interim storage facility, and a geologic repository. The Strategy fully endorses the need for a consent-based process for siting facilities and highlights the need for both a new waste management and disposal organization and mechanisms to assure adequate and timely funding. The Administration continues to believe that these elements are necessary to provide the stability, focus, and credibility to build public trust and confidence and to assure overall success of the nuclear waste mission.

The Administration, through NE, is undertaking activities within its existing authority to plan for the eventual transportation, storage, and disposal of used nuclear fuel. To support the evolution of the domestic UNF inventory, special emphasis is placed on confirmatory research to provide further understanding of the long-term behavior of high burnup fuels.

NE's Used Nuclear Fuel Disposition subprogram conducts scientific research and technology development to enable storage, transportation, and disposal of UNF and wastes generated by existing and future nuclear fuel cycles.

For disposal research and development, activities continue to further the understanding of long-term performance of disposal systems in three main geologic rock types: clay/shale, salt, and crystalline rock. In addition to focusing on mined geologic repositories, the Department is evaluating the possibility of disposal of certain radioactive waste in deep boreholes. The Department has initiated a process to identify a volunteer site for the core research and development activity related to the deep borehole (DBH) disposal concept, the DBH Field Test. t.

For Storage and Transportation research and development, because of the evolution of the domestic UNF inventory, special emphasis is placed on confirming and increasing our understanding the behavior of high-burnup fuels. In addition to laboratory testing, modeling, and observations at existing storage installations, a key R&D activity is the Full-Scale Storage Cask Demonstration, in the process of being implemented on a cost-share basis at a commercial nuclear utility. This Demonstration will be beneficial by 1) benchmarking the predictive models and empirical conclusions developed from short-term laboratory testing and 2) building further confidence in the ability to predict the performance of these systems over extended time periods.

Another important initiative within NE involves the development of accident tolerant fuels, a next-generation nuclear fuel with higher performance and greater tolerance for extreme, beyond design basis events. . These fuels would give operators additional time to respond to unforeseen conditions, such as those experienced at Fukushima-Daiichi. The program is framed on a three phase approach from feasibility to qualification to commercialization and is executed through strong partnerships with national laboratories, universities, and the nuclear industry. The industrial research teams, led by AREVA, Westinghouse, and General Electric, plan to begin irradiation of their proposed fuels in the Idaho National Laboratory (INL) Advanced Test Reactor next year. DOE has also expanded collaboration with our international partners, including France, Japan, and the United Kingdom, as well as multi-lateral programs under the leadership of the OECD-Nuclear Energy Agency and the International Atomic Energy Agency.

Investing in Research and Development Infrastructure

Research, development, and demonstration programs are dependent on an infrastructure of experimental facilities, computational facilities, and highly trained scientists and engineers dedicated to meeting the needs of the Nation. The Nation's nuclear research, development, and demonstration infrastructure incorporates a broad range of facilities, from small-scale laboratories to hot cells and test reactors, up to full prototype demonstrations. Computing facilities ranging from desktop workstations to highly parallel supercomputers at the national laboratories are routinely employed to gain new insights and guide experiment design. The high cost of creating and maintaining physical infrastructure for nuclear research, development, and demonstration, including the necessary safety and security infrastructure, requires close alignment of infrastructure planning with programmatic needs to ensure capabilities are planned, maintained and available to support NE missions. To enable and facilitate research and development activities, NE's Idaho Facility Management program maximizes the utility of existing facilities and capabilities through focused sustainment activities and cost-effective rehabilitation. Activities focus on safe and compliant operation of the Idaho National Lab's (INL) nuclear research reactor and non-reactor research facilities, while conducting corrective and cost-effective preventative maintenance activities necessary to sustain this core infrastructure.

In spite of these efforts to maximize the effectiveness of the established infrastructure, additional investments are needed to achieve further progress in advanced nuclear technologies. In its FY 2015 budget request, NE proposed to focus investments on reestablishing a domestic transient testing capability with the TREAT reactor at INL. This capability will enable the NE research and development programs to understand fuel performance phenomenology at the millisecond to second time scale as well as provide a capability to screen advanced fuel concepts, including accident tolerant fuels, which allows for early identification of the limits of fuel performance.

Moving forward, the Department will continue to assess infrastructure capabilities to ensure the U.S. maintains a key leadership position in the international development of advanced nuclear technologies. By staying at the forefront of nuclear technology, we are able to ensure that the U.S. safety and non-proliferation standards are adopted internationally while providing clean, affordable, baseload power globally.

Conclusion

In conclusion, the programs of the Office of Nuclear Energy support the many aspects of this important energy source, from reactors, to used fuel management, to infrastructure. NE's programs strive to ensure both the current fleet and advanced light-water technologies are available to meet the Nation's energy security and clean energy needs. As we look further into the future, we recognize that reactors using coolants other than water appear to offer important attributes. NE supports research on these advanced reactor concepts and seeks international cooperation through the Generation IV International Forum. The current generation of modern plants and the development of the passively safe SMR designs, both using light water coolant, serve as a vital bridge between the reactors of today and the potential for non-light water reactors in the future. In the United States we now have immense knowledge of light water reactor systems and can design and regulate them with the highest confidence for safe operations. Research today should focus on providing that level of confidence for these new concepts for tomorrow.

Dr. Peter B. Lyons was confirmed as the Assistant Secretary for Nuclear Energy on April 14, 2011 after serving as the Acting Assistant Secretary since November 2010. Dr. Lyons was appointed to his previous role as Principal Deputy Assistant Secretary of the Office of Nuclear Energy (NE) in September 2009.

Under Dr. Lyons' leadership, the Office has made great strides in incorporating modeling and simulation into all programs through the Nuclear Energy Advanced Modeling and Simulation program and the Energy Innovation Hub. He focused on management of used fuel by contributing to the development of the Administration's *Strategy for the Management and Disposal of Used Nuclear Fuel and High-Level Radioactive Waste*. In addition, NE established the Small Modular Reactor Licensing Technical Support program for a new generation of safe, reliable, low-carbon nuclear energy technology. And he championed the Nuclear Energy University Program, which has successfully supported U.S. universities in preparing the next generation of nuclear engineering leaders.

Prior to joining DOE, Dr. Lyons was sworn in as a Commissioner of the Nuclear Regulatory Commission on January 25, 2005 and served until his term ended on June 30, 2009. At the NRC, Dr. Lyons focused on the safety of operating reactors, even as new reactor licensing and possible construction emerged. He was a consistent voice for improving partnerships with international regulatory agencies. He emphasized active and forward-looking research programs to support sound regulatory decisions, address current issues and anticipate future ones. He was also a strong proponent of science and technology education.

Before becoming a Commissioner, Dr. Lyons served as Science Advisor on the staff of U.S. Senator Pete Domenici and the Senate Committee on Energy and Natural Resources where he focused on military and civilian uses of nuclear technology from 1997 to 2005. From 1969 to 1996, Dr. Lyons worked at the Los Alamos National Laboratory where he served as Director for Industrial Partnerships, Deputy Associate Director for Energy and Environment, and Deputy Associate Director-Defense Research and Applications. While at Los Alamos, he spent over a decade supporting nuclear test diagnostics.

Dr. Lyons has published more than 100 technical papers, holds three patents related to fiber optics and plasma diagnostics, and served as chairman of the NATO Nuclear Effects Task Group for five years. He received his doctorate in nuclear astrophysics from the California Institute of Technology in 1969 and earned his undergraduate degree in physics and mathematics from the University of Arizona in 1964. Dr. Lyons is a Fellow of the American Nuclear Society, a Fellow of the American Physical Society, and was elected to 16 years on the Los Alamos School Board.

Dr. Lyons grew up in Nevada and is a resident of Washington, DC.

Chairwoman LUMMIS. I thank the witness for his testimony.

I will remind the Members that Committee rules limit questions to five minutes and the Chair at this point will open the questioning.

Now, I want to set this up, Dr. Lyons, to see if I understand this correctly. The Energy Policy Act of 2005 established the Next Generation Nuclear Plant project, and as I understand that, it was to develop a prototype reactor for an eventual hybrid energy system. And it was supposed to be accomplished through cost-shared R&D, as well as design, construction, and operation on behalf of the Alliance. Now, is that a quick summary, an accurate summary of the NextGen Nuclear Plant project?

Mr. LYONS. Yes.

Chairwoman LUMMIS. Okay. Then, it is also my understanding that DOE and the NextGen Nuclear Plant Alliance have reached somewhat of an impasse over cost-shared distribution, that the Alliance is asking the Department to frontload its portion of cost-share while DOE maintains a cost-share at 50/50 throughout going from day one until it comes online.

I understand also that there have been some successes thus far in this program, including the development of TRISO fuel, so I am very interested in knowing is there an impediment for DOE to exercise its authority to host, for example, private development of prototype reactors at a DOE site?

Mr. LYONS. Thank you for those questions, all very good ones. And there are several different questions inherent in what you just asked.

Your description of NGNP I believe is accurate. Very successful R&D has been conducted on that program. You mentioned TRISO fuel. We now have very high confidence in our ability to produce TRISO fuel to the highest standards and TRISO fuel is capable of withstanding extremely high temperatures in any accident scenario. It is an incredibly robust type of fuel which we believe can have applications in many, many future systems. There are other areas of strong research for NGNP. I am sorry, other areas of strong research that were conducted as part of NGNP.

Now, you are also quite right that as we came to the point in our R&D programs where it became feasible to look towards actually moving ahead with a demonstration reactor, we asked our advisory committee to evaluate the status of the research. At the same time we were discussing with the NGNP alliance their interest in moving ahead.

I might note that when we started the NGNP program, there were a number of studies which pointed out the cost efficiency of this approach for gas greater than $8, and when we started NGNP—and I have to admit I was one of the co-authors of the language—when we started that program, gas was way over $8.

Chairwoman LUMMIS. Yeah.

Mr. LYONS. At the time we got close to being able to move towards a developmental program; that was definitely not the case.

Chairwoman LUMMIS. Um-hum.

Mr. LYONS. And while, when we initially wrote the language, there was very strong interest from industry in looking towards the 50/50 cost-share, as we moved towards the point in time when we could have started into development, as you described correctly, their interest was in the DOE frontloading the expenses and they might—they would pick it up later if they deemed appropriate. That is not my understanding of the Congressional intent on cost-sharing as written in EPAct '05, and furthermore, to accomplish what they would have suggested would have taken essentially the entire R&D budget of my office simply for NGNP.

So we have continued the research on the TRISO fuels, the graphites, we have continued to work with and even to some extent support the NGNP Alliance as they look towards possible opportunities in the future. But this question of what is the appropriate cost-share certainly could be the subject of more discussion within Congress and exactly how the intent was to formulate that cost-share, but in my mind, it is important to have strong industry support, as evidenced by cost-share, before one moves ahead to actually build a prototype reactor of any of these.

You also, right at the end, asked the possibility of utilizing—I think you said DOE sites in moving ahead with advanced reactors. I think that is also a subject of great interest. We can certainly discuss it further. I am not aware of any fundamental impediments to that. There would be a number of challenges and I think we could talk through what those challenges might be if you wish.

Chairwoman LUMMIS. I would like to go there. Can you tell— oops, my time is expired. That went really fast.

Chairman SMITH. You can have more time if you want.

Chairwoman LUMMIS. Oh, well, thank you, Mr. Chairman.

I do think that I will recognize Ms. Bonamici for five minutes.

Ms. BONAMICI. Thank you very much, Madam Chairwoman, and thank you for allowing me to join you this morning even though I am on the full Committee, not on this Subcommittee. I wanted to be with you today, especially because of Mr. McGough from NuScale. But thank you so much, Dr. Lyons, for being here.

I wanted to ask you, the Energy Policy Act of 2005 created the— of course the Next Generation Nuclear Plant project along with timelines for completing each of the project's three phases. Apparently, there have been some barriers that have arisen. Can you just talk a little bit about the reasons for the delays in that project and then I want to save time for another question as well, please.

Mr. LYONS. Thank you for the question. I think I tried to address some of that on the previous questions.

I think the research on NGNP has gone extremely well. We have made dramatic progress. But as I indicated, at the time NGNP was formulated, there were many studies saying $8 gas was the break-even point. We don't have $8 gas today and we are way below the breakeven point.

There—I also just alluded to the I would say difference in opinion between the NGNP Alliance and me, my office, on what it means to cost-share. Their proposal was that we construct the reactor and they would—and I am paraphrasing this greatly—but that we would construct the reactor and that they would decide later if they wished to build the actual systems, operational systems, and that over the long run one would achieve a 50/50 cost-share.

My understanding of EPAct—I think it is Section 988 perhaps of EPAct—was that a cost-share means a continual cost-share over the life of the program. Now, that could be subject to interpretation and certainly for evaluation by Congress. I hope that is at least a bit of an answer.

Ms. BONAMICI. Yes, thank you for expanding on that.

And then Mr. Swalwell in his opening remarks talked about the issue of safety, which of course our constituents are concerned about as well. And I represent a district out in the northwest where we spend a lot of time talking about resilience and what will happen. We are—we have the Cascadia Subduction Zone off our coastline and we are having a lot of conversations about how we deal with the eventual earthquake and tsunami.

So can you talk a little bit about the lessons that the Department has learned from the Fukushima disaster, what work is being done, not necessarily just in siting but in structure, to make sure that there is that preparation for sites in areas like the Northwest where there will eventually be earthquakes and tsunamis?

Mr. LYONS. Thank you for that question as well.

If I were to start with the lessons—well, I could talk for days on the lessons of Fukushima. However, if I were to start with the single most important lesson it was on the importance of having an independent regulator. They—Japan did not have an independent regulator like the NRC. While I was at the NRC, there were many advances that were made in U.S. plants, for example, to prevent— to respond to a station blackout. We shared that information with the Japanese regulator at that time. The Japanese regulator did not elect to make those requirements on Japanese plants. Japan has now moved to an independent regulator away from their previous system where their regulator was part of METI and the I in METI is industry. So, number one lesson, have an independent regulator. We have one and I was proud to serve with the NRC.

In terms of lessons from the actual events at Fukushima, certainly the NRC has evaluated those but I think it is also fair to say that our plants are extremely well prepared because of any number of requirements that we have required—that we had demanded of the nuclear industry.

But specific areas of research on which we have—what we have expanded post-Fukushima, one would be so-called accident-tolerant fuels. The current generation of fuel systems use a zirconium cladding. Under accident conditions that creates hydrogen. When you have too much hydrogen, things blow up and there were—and that took a very bad day at Fukushima Daiichi into an absolute crisis.

We believe it is possible to generate—and to come up with a new generation of fuel systems that would greatly minimize the production of hydrogen under an accident scenario. That has been very well supported in Congress, about 60 million a year. We have been making dramatic progress.

Introducing a new fuel system is a big deal in the nuclear industry and that is going to take more than a decade to do this but we are making good progress. There are good ideas and it is my hope that we will start testing probably in 2018 on the initial—we will make some down-selects in 2016. We will have the first testing— trenchant testing in 2018 for accident-tolerant fuels. And if we can develop that, that will be another significant step forward.

24

But on other points—and I am sorry I am probably taking too much time here—you mentioned an interest in NuScale in the small module reactors. The fact that NuScale and any of the SMRs are—that we are interested in—are sited underground, it gives them substantially more seismic resistance. In addition, the design of the NuScale plant increases—it is a long word, the probabilistic risk assessment of the plant, the probability of an accident is many decades lower in the NuScale design. That plant is dramatically safer than our existing plants, which are already very safe. And again——

Ms. BONAMICI. Terrific. I look forward to hearing from Mr. McGough about that.

My time is expired and thank you, Madam Chairwoman.

Chairwoman LUMMIS. I thank the gentlelady and yield to the gentleman from Texas, Mr. Smith.

Chairman SMITH. Okay. Thank you, Madam Chair.

Let me continue following up on that same subject that was mentioned by the Chair a few minutes ago, that is to say the cost, but on the way there, Dr. Lyons, thank you for your encouraging remarks. You are being very positive. You are talking about how we can improve things for the future and that is what this hearing is all about.

We have been told that sometimes it costs up to $1 billion to get through our NRC's licensing process. Is there any way to reduce the cost of that? I have several questions. Is there any way to reduce the cost? Is there any way to streamline the process? Is there any way to make it easier to construct safe nuclear reactors in the future?

Mr. LYONS. Thank you for that question, and again, there are many ways I can answer that question.

The billion-dollar number is frequently used but let's remember that that billion is far more than just what—than just the actual work done by the NRC. In order—under Part 52, the new licensing approach at the NRC, the vendor—and NuScale if you want to use them as an example—has to prepare a very complete design. They have to go way into the engineering details of the plant in order to answer all of the questions from the NRC under Part 52. Now, under Part 52 the goal is that you end up with a certified design, and once you have that certified design, as long as you stay with it, then you don't go back through the safety analysis again. And that is believed to be a very effective way of advancing nuclear power in this country.

Chairman SMITH. So part of the cost is the design which they would have to do anyway, is that what you are saying?

Mr. LYONS. Yes.

Chairman SMITH. Okay.

Mr. LYONS. And I think it is fair that we talk about a billion that we recognize——

Chairman SMITH. Okay.

Mr. LYONS. —that that is included but at the end of the game you end up with success with a certified design but then you can take to any site in the country and not go back through the safety analysis.

Chairman SMITH. Okay. And let me follow up again on those other questions. I think it takes close to, what, five years to get through the regulatory process now? I am not sure how many years but that is what I have read but how can we expedite the process? It just seems to me that if we are trying to accomplish the good designs, if we are trying to increase nuclear energy, there is bound to be a way to try to actually encourage companies to go that direction and to not unduly prolong the process.

Mr. LYONS. As the NRC has evaluated the SMR designs, they have published a schedule of 39 months that they intend to follow once an application is filed. That remains to be tested and I am certainly hoping they will succeed.

You asked what could possibly be done to improve that. I would note that NuScale, as one of the SMRs that we are supporting, has taken advantage of the so-called pre-licensing process in which they can submit white papers to the NRC on specific design aspects and gain comments back from the NRC. I think that is a very effective way of perfecting a design.

If you ask me for one possible improvement in that, right now, the NRC delivers an informal opinion on those white papers. I think one could imagine that it could be even more useful to companies like NuScale or mPower if it was a formal decision, and that might be a question that you address through NuScale——

Chairman SMITH. How much time would that save?

Mr. LYONS. Well, I think what it would do would be to provide far more confidence to a vender that what they have submitted in a white paper and had some pre-analysis is actually going to be accepted. Right now, there is 30 or more white papers that are put in, they have got comments on all of them, but they don't have the confidence that there won't be a change later when the commission evaluates it.

Again, this is just a suggestion and it certainly would be appropriate to discuss with NuScale and with the NRC.

Chairman SMITH. Okay. And is this a subject that you are discussing with NRC? You have some control over what they do. Are you encouraging them to expedite the process and reduce the cost if it is possible?

Mr. LYONS. Well, we continue to have frequent interactions with the NRC on all of our programs and keep them as informed as we can on the directions that we are going——

Chairman SMITH. Yeah.

Mr. LYONS. —and asking how we—how our research can help them.

Chairman SMITH. Yeah. I guess I am looking to see if you will go beyond just keeping them informed and actually try to spur them to take some steps to reduce the cost and the time involved with the licensing process.

Mr. LYONS. The reason I am giving you a delicate answer is they are an independent agency. They highly value their independence. I know that having served there. And I think one probably wants to be a little bit judicious in how strongly one makes suggestions that could be interpreted as undermining their independence.

And I just made the comment, too, about the importance of an independent regulator to avoid a Fukushima, too, so it is important.

Chairman SMITH. Independent agencies need oversight and suggestions as well so——

Mr. LYONS. And certainly that is—yes.

Chairman SMITH. Okay. Thank you, Dr. Lyons.

Thank you, Madam Chair.

Chairwoman LUMMIS. The Chair now recognizes another Member of the Texas delegation, Mr. Veasey.

Mr. VEASEY. Thank you, Madam Chair.

I wanted to talk with you a little bit about the stages for licensing advanced reactors. We have received testimony that there should be stages of review by the NRC similar to how the FDA has three phases of review before a new drug is allowed to go out onto the marketplace. The argument is that this would provide much earlier and a clearer signal to investors that a new nuclear technology is meeting or failing criteria set by the NRC. As a former NRC Commissioner do you have any thoughts on this?

Mr. LYONS. Well, thank you for that question and I would go back to my comment of just a few minutes ago on the pre-licensing reviews. I think those are an extremely effective way of a company testing the waters if you will, starting to raise appropriate questions with the NRC, and gaining feedback from the staff.

Now, this—also the suggestion I made just a minute ago was I think it would be worth discussing with the NRC whether to carry it out a little bit further and instead of just a staff opinion that comes back maybe asking if that opinion have a little bit more weight so that the—a potential vendor would have more confidence that if they stay with a particular design aspect that it will be accepted by the Commission later on. So I think that is at least an approach towards the question you have asked.

Mr. VEASEY. Thank you very much. As far as the commercialization of advanced reactors again, what do you believe are the necessary components of a public-private partnership that can ultimately take these advanced reactors from the lab to the marketplace?

Mr. LYONS. At least one example has been the NP 2010 Program, highly successful program that Congress supported for a number of years which was cost-shared with both Westinghouse and GE and has led to certified designs. We are trying to do exactly the same thing with the small modular reactors. We may get to the point where there is sufficient maturity of some of the non-light water designs that it would be—that Congress might want to consider that in the future.

In any case, my point would be that the cost-shared—that the cost-shared work towards design certification has already proven to be very successful. In the case of NP 2010 in the case of the SMRs, at present we have not asked for support beyond the design certification anticipating that at that point there is sufficient information for industry to make their own decisions and, of course, the loan guarantee program that the Department has also consists in this.

Whether one can go still further, one could if that were deemed appropriate from a Congressional standpoint, and certainly cost would escalate appropriately.

Mr. VEASEY. What about steps that commercial vendors and utilities need to take to ensure their ability to accept advanced reactor technology as part of their energy portfolio?

Mr. LYONS. I think part of that answer will be addressed by Dan Lipman in his testimony on the next panel. I found his testimony very interesting where he notes that industry has formed a working group now devoted to advanced reactor technologies. That is somewhat analogous to what they did during the time when we were working on design certification of the Westinghouse and GE reactors. They have a similar model that they are using now on the small modular reactors and I think in general having industry organize to explore their own interests in any particular reactor design is highly advantageous.

Mr. VEASEY. Madam Chair, thank you very much. I yield back my time.

Chairwoman LUMMIS. I thank the gentleman and recognize the gentleman from Texas, Mr. Weber.

Mr. WEBER. Thank you.

Dr. Lyons, as you know, the United States has engaged in ongoing climate negotiations with the U.N., final agreement expected in Paris 2015. Do you believe that the benefits of nuclear power should be specifically recognized in the UNFCCC agreement to be reached in Paris?

Mr. LYONS. I guess I would answer, Mr. Weber, that certainly the benefits of clean energy need to be recognized. Whether nuclear needs to be called out specifically in that, I don't have an opinion. There are many, many studies showing that in order to achieve the clean energy desired in the future that nuclear will be a significant part.

Mr. WEBER. Do you know of any other energy as reliable and capable of producing the kind of megawatts necessary as nuclear energy?

Mr. LYONS. Well, that is why any of the studies that I am referencing, for example, the recent World Energy Outlook from the International—IEA, International Energy Agency, notes the importance of a strong nuclear component looking into the future for exactly that reason.

Mr. WEBER. So that is kind of a roundabout way of saying yes?

Mr. LYONS. Well, again, I am saying I don't think that you can achieve what we need without nuclear but I also think——

Mr. WEBER. But I mean the importance of recognizing it in the agreement reached in Paris?

Mr. LYONS. The only reason I am hesitating, sir, is that different communities, different regions, different countries are going to have different mixes of power appropriate to whatever their situation is.

Mr. WEBER. Yeah, but let's focus——

Mr. LYONS. So I don't want to——

Mr. WEBER. Let's focus on the United States, though.

Do you think emissions—let me back up. What role will nuclear play in meeting global emissions targets that we can expect in the agreement in Paris? I mean we have already talked about 60 percent of the energy would—basically zero emissions so would you expand on that a little bit for us?

Mr. LYONS. Well, I would just again note that any study I have seen—I happened to reference the World Energy Outlook that was published just recently—certainly notes that nuclear is going to have to play a strong role——

Mr. WEBER. Okay.

Mr. LYONS. —as we look forward into—to reach the goals that are in—that are——

Mr. WEBER. Sure.

Mr. LYONS. —required.

Mr. WEBER. So that is to say that you don't believe those targets could be reached without nuclear when you say it has to play a strong role?

Mr. LYONS. That is—I believe that is true and that is consistent with the President's all-of-the-above strategy.

Mr. WEBER. Do you know if the new Green Climate Fund for international mitigation efforts can be used to support nuclear projects?

Mr. LYONS. I do not know, sir. .

Mr. WEBER. Okay. Do you believe it should be used to support nuclear projects? Noting your earlier comments, you can't reach those targets without nuclear.

Mr. LYONS. I know so little about that fund that I hesitate, sir.
.

Mr. WEBER. Yeah.

Mr. LYONS. I mean in general, yes, I think nuclear should be recognized for its clean energy—.

Mr. WEBER. But you do get paid to work—you do get the money, right, so you know a little bit about money, and so if we have got dollars being spent for Green Climate Fund and nuclear is—you said you can't reach that target without nuclear, doesn't that make sense that some of that fund should perhaps be used to support nuclear projects?

Mr. LYONS. Again, I am not sufficient—nuclear is going to be a part of a future solution. I don't know enough about that fund to give you a credible answer, sir. .

Mr. WEBER. Well, it is dollars for Green Climate Fund and it seems like we ought to be including nuclear in that.

Let's change over to Yucca Mountain for a second. In a letter to NRC Chairman Macfarlane dated November 18, 2014, from Senator Patty Murray, she stated ''Over the last 30 years, independent studies have pointed to Yucca Mountain as the nation's best option for a nuclear repository for high-level waste. At the same time, Congress in every previous Administration have voted for, funded, and supported pursuing this option. The recent completion of Volume 3 of SER reaffirms that Yucca Mountain is the right solution for the United States.'' Do you agree with Senator Murray?

Mr. LYONS. The fact that the SER Volume 3 stated the safety from a post-closure standpoint of Yucca Mountain is not a surprise. The Department submitted our application for Yucca Mountain in 2008. It doesn't change the fact that we believe—I believe strongly that Yucca Mountain is not a workable solution and I would be happy to go into that in as much detail as you would like. I have spent a lot of my career involved with Yucca Mountain.

Mr. WEBER. Are you familiar with the Waste Control Specialists site low-level radiation out in Andrews, Texas?

Mr. LYONS. I have visited them multiple times.

Mr. WEBER. Okay. What is the difference between a low-level radiation and high-level radiation, notwithstanding the obvious? When you use fuel rods, for example, we expend most of them and I understand France has a method for reclaiming a lot of that energy—what is the difference between low-level waste and high-level waste? Can you expand on that? And I am out of time but if you can do that quickly.

Mr. LYONS. In this country we have a definition of high-level waste that refers to its use in a reactor and ties it to used fuel. However, the—a simpler definition is simply as you said; it is kind of obvious. One is low and one is high-level radiation. As far as France—and we could talk about France a great deal if you want in the future——

Mr. WEBER. But when you use fuel in nuclear reactors, do we not try to use up all of that energy and all of that fuel?

Mr. LYONS. With our current generation of light water reactors we come nowhere close to using the full energy content of the fuel resource.

Mr. WEBER. Okay.

Mr. LYONS. That is one of the advantages of advanced reactors.

Mr. WEBER. Okay. Thank you. I yield back. Thank you.

Chairwoman LUMMIS. I thank the gentleman.

The Chair recognizes the gentleman from California, Mr. Swalwell.

Mr. SWALWELL. Thank you, Madam Chair.

And, Mr. Lyons, as you know, I represent Lawrence Livermore National Laboratory and Sandia National Laboratory in Livermore, California, and while these labs do not work directly on nuclear energy, I know that Sandia, for example, played an important role in the accident response and after-accident analysis with respect to Fukushima. And I was hoping you could tell us about the role our national labs are playing to keep us safe, as well as how important it is that our national laboratory system advances science, and how DOE is utilizing our national labs to ensure that we are doing everything we can to keep existing reactors safe.

Mr. LYONS. Thank you, Mr. Swalwell.

There is no question that our national laboratories are an incredible national resource for science and technology. I make extensive use of all of the national laboratories in my program. Livermore is not one of the predominant ones. Sandia is. You mentioned Sandia's very strong capabilities in severe accident management and a number of other areas that are ongoing at Sandia and I have extensive funding at Sandia.

But in general the national labs are a vital resource and I make extensive use of them.

Mr. SWALWELL. Great. And, Mr. Lyons, I and many people in my district have strong concerns about the safety of nuclear energy and I was hoping you could tell us about the research that the Department of Energy is conducting to improve existing technology and make our nuclear plants safe.

Mr. LYONS. The—I would first start with the Accident-Tolerant Fuel Program that I mentioned a little bit earlier. The national labs, industry, and universities are heavily involved in the Accident-Tolerant Fuels Program where we are seeking to develop a class of fuels that would ideally not emit hydrogen, at least emit far less hydrogen in an accident scenario. And it was the hydrogen explosions at Fukushima Daiichi that took a very bad situation into a catastrophe.

Mr. SWALWELL. And, Mr. Lyons, could you talk more about Sandia's nuclear accident modeling software MELCOR?

Mr. LYONS. MELCOR has been vital throughout the industry. I was over at—in Tokyo within days of Fukushima, as were leaders from the Sandia Severe Accident Program. MELCOR was used extensively post-Fukushima. We continue to use MELCOR. MELCOR is the prime code used in this country for severe accidents and both we and the NRC use it extensively.

Mr. SWALWELL. Great. Thank you, Dr. Lyons, and I yield back the balance of my time.

Chairwoman LUMMIS. I thank the gentleman.

The Chair now recognizes the gentleman from Kentucky, Mr. Massie.

Mr. MASSIE. Madam Chair, I just want to say thank you and thank you for letting me serve on your Subcommittee. It has been a pleasure serving here for you and with you on the Oversight Committee. I will note that one of your questions went viral on YouTube the other day on Oversight. If only we could get something in the Energy Subcommittee to go viral. Maybe we need to bring a working reactor here for that to happen. That might get it done.

I have always been a supportive of an all-of-the-above energy plan both in public life and in my private life. I started a company right off the MIT campus and the directions to our company were take a right at the nuclear reactor and a left at the candy factory. So we were never worried about the safety of that.

I was astounded; I looked at the cost of that nuclear reactor. It only cost $3 million to build that in 1956. Yeah. But even accounting for inflation that was astounding and there is something wrong about the price of nuclear energy right now, the fact that it costs so much to build a new plant these days.

You know, in my personal life I drive a Tesla. It is an electric car but it has Friends of Coal license plates just so people know where the energy comes from. To balance out the karma there I live off the grid. I just added 3 kilowatts of solar panels to my 10 kilowatt array, and I used to work in an oil refinery.

But I find it disturbing that the State of Kentucky has a moratorium on building any nuclear plants and I think that is a problem, I think it is wrong, and I think it is to the detriment of the citizens of Kentucky unfortunately.

But, there is sort of public opinion about this and the elephant in the room here is what do we do with the nuclear waste? And I think Mr. Weber asked you earlier about Yucca Mountain and that is sort of my question. We see trucks going through the district that have canisters that have nuclear material in them. What are we going to do about the nuclear waste? If the answer is not Yucca Mountain, what is the answer?

Mr. LYONS. The Administration published a strategy in January of 2013 based on the recommendations of the Blue Ribbon Commission.

Mr. MASSIE. What is your recommendation?

Mr. LYONS. I think the Blue Ribbon Commission had fabulous suggestions, to move ahead with a consent-based process. And I think that if there was a legislative basis to move ahead with a consent-based process——

Mr. MASSIE. What does that mean for my constituents back home, consent-based process? What is the answer is what we want to know? And I only have two minutes and I know it is more complicated than that.

Mr. LYONS. I think consent basis means frankly exactly the opposite of the Yucca Mountain situation. I grew up in Nevada, worked in Nevada, worked at the test site, worked with Yucca Mountain, directed the research on Yucca Mountain. I know it rather well. But I also am well aware that the Nuclear Policy Waste Act amendments of '87 are viewed in Nevada as the ''screw Nevada'' bill. There was never a consent basis at—in Nevada for the—for Yucca Mountain and it has led to a rather—to say it is polarized is putting it mildly.

Mr. MASSIE. If the——

Mr. LYONS. On a consent basis we would avoid that.

Mr. MASSIE. If nuclear energy is going to flourish and remain a viable option for us, we have to solve this problem. Do we have time to do this consent-based process? Or, I am sorry, consensus?

Mr. LYONS. Well, I would submit that if we don't do a consent-based process, we will have rate difficulties ever succeeding. So the—we—the current storage of the used fuel in pools and dry casks is safe but there is no question that that is not a long-term solution. And eventually we need to move, I believe, as the BR—Blue Ribbon Commission said, to centralized interim storage and to a repository but doing it on a consent basis. And there are a number of communities, in some cases even states that have expressed interest in being considered for housing such facilities. And I think if we tried on a consent basis, discussed how this could be done with the utmost attention to safety, I think we could succeed.

Mr. MASSIE. I think we have to succeed with doing something with the waste because this is the elephant in the room, what are you going to do with it? And this is what drives public opinion I think. I am convinced that the safety issue has been solved but clearly dealing with the waste has not been solved.

Let me switch to a lighter topic because I would be remiss if you were here and I didn't get a chance to ask this question for my constituents who are always asking me about this. Does thorium have a place in our nuclear future?

Mr. LYONS. We have evaluated thorium-based cycles many, many times. Given that we have made a massive commitment in this country to a uranium-based cycle, I see no compelling reason to move towards a thorium cycle. I—

Mr. MASSIE. If we weren't so heavily invested though in this path, does it make sense? I mean does it make sense in other countries that——

Mr. LYONS. If you are starting from scratch, I think one could make a decision to go either way but some of the claimed advantages for thorium, which I hear frequently from people who would like us to put more money into thorium such as that it is proliferation-resistant, are simply false. There was a recent report done by the Nuclear Energy Agency of the OECD on thorium systems which certainly made this point, that they are anything but proliferation-resistant.

Can you make them work? Yes, you can make them work. Is there an advantage to doing it? I haven't seen it.

Mr. MASSIE. Thank you very much. I—my time is expired.

Chairwoman LUMMIS. I thank the gentleman, recognize the gentleman from Texas, Mr. Neugebauer.

Mr. NEUGEBAUER. Thank you, Madam Chairman.

Dr. Lyons, according to recent reports from some of the NGO organizations that there is a significant chance that nuclear power plants may close as a result of low natural gas prices and renewable energy subsidies, to what extent has DEO—DOE assessed those scenarios?

Mr. LYONS. We have studied that, sir, in considerable detail. We are very concerned that the closure of any clean energy resource in the nation only complicates our eventual quest for a—an overall clean energy system. As we have evaluated the reasons for some of those closures and some of the economic pressures, we have yet to identify a federal lever that could be used to protect those plans. Most of what can be done is on a State basis and there are widely publicized negotiations going on in a number of States, certainly Illinois and New York would be two very prominent, where there are negotiations at the State level that might involve power purchase agreements as one example in order to keep marginally economic nuclear power plants online under the current market system.

There also are efforts, for example, in PJM region to move towards a so-called capacity auction that would do a better job of valuing the attributes of nuclear, that it is always there, very reliable, highly resilient, very important in maintaining overall good stability. Those are not attributes that are currently valued as perhaps they might be and it is—these questions of how the markets value nuclear power remains a complex issue but I think that PJM in their region of the country are starting to ask these very important questions.

Mr. NEUGEBAUER. So I guess one of the questions—and do we have conflicting policy in some way where we are subsidizing other renewables and so making it difficult to actually have price discovery of what is the market, for example, power because we are distorting that in some ways with some of these subsidies?

Mr. LYONS. Well, I think it is fair to note that there is a number of different factors that are entering into these questions. Certainly low natural gas prices, while a tremendous boon for the country, also are at least challenging to any of the clean energy systems. There is also flat or decreasing electrical demand in many parts of our country as more and more efficiency measures are coming into play. That, too, makes it very difficult, so there is a number of different stresses on the nuclear power plants. And particularly for the relatively small single unit sites, they are the ones under the greatest stress and those are the ones that, as I indicated, that we have been discussing whether there is a direct federal action and we haven't found it yet. We are still—if we find it, that would certainly be interesting.

Mr. NEUGEBAUER. So on one of the things that I guess—and I am always a little reluctant to point out France but one of the things that they have done over the years, they have a pretty robust nuclear presence over there and one of our opportunities to sit down with some of the people over there—and of course they recycle a lot of their nuclear waste and to the point where they—as I understand it—I am—don't understand all of the science of it but they reprocess a lot of the—and what they basically said is that we keep reprocessing and reprocessing and reprocessing and so the volume ultimately that we dispose is much smaller. Is that something that the United States should be thinking about?

Mr. LYONS. Well, first, we have robust programs looking at R&D on advanced reprocessing.

Reprocessing certainly opens many questions, including nonproliferation and environmental ones. It is fair to say that the type of reprocessing that is done in France at La Hague would not be licensed in the United States with the level of emissions that they have. It also would be somewhat misleading to say that they reprocess over and over. They reprocess once, go to MOX fuel, and then they are storing the MOX fuel.

Now, their eventual goal is to move towards fast reactors and closing the fuel cycle. They are a long ways from doing that but they are going at least one step of reprocessing and at least—I would—I think I have made the point that from—our concerns would be both from an environmental and nonproliferation standpoint, which is why we have the research programs to continue to evaluate options looking into the future. And I think the country may at some point want to evaluate whether they want to move towards a closed cycle, but in my mind that would be made after one has demonstrated a repository because you need a repository whether it is an open or a closed cycle. France still needs a repository. They are building one.

Chairwoman LUMMIS. I thank the gentleman.

And the Chair without objection will recognize Mr. Rohrabacher from California for five minutes.

Mr. ROHRABACHER. Thank you very much, Madam Chairman.

And let me congratulate you on a tour of duty here on this Committee and you have done us proud and served your country well with the leadership you have provided, and we wish you all the luck and we will be working with you on your new assignment as well.

So I am a bit disturbed by some of the directions that we are talking about today and I know we have had this exchange before and it just seems to me that when we talk about the development of Next Generation Nuclear Power Plants and as we are stepping forward, the words light water reactor continue to be part of the game. And I have been told by numerous engineers, renowned engineers, people who know what they are doing and—who tell me that we now are capable of building nuclear power plants, for example, General Atomics has a plan for a high temperature gascooled reactor that can be done and that we have small modular reactors because—various sizes, but yet—and that reactor would be—and I am sure—and I have talked to other scientists and engineers about other approaches and they—I don't know—you suggested that—this—oh, this thing about thorium is that some of the claims are not true, but there are a number of approaches that I have been told would eliminate the leftover waste problem, which is a huge challenge for us to overcome before the public is going to accept further investment into nuclear energy.

But every time I hear about—coming back—what will be built, again, it is light water reactors. So this money that you are talking about now being expended will go to light water reactors which have some of the same defects that we have experienced with Fushimora? I guess I am not pronouncing it right—Fukushima. And I don't understand what is going on here. Why are we spending money to build basically reactors based on the same concept that Fukushima was built on and that we have been building ever since World War II?

Mr. LYONS. Well, thank you for the question, sir, and there certainly could be many answers to that.

The reactors that we are—that we have supported through the NP 2010 program or that we are supporting through the SMR program——

Mr. ROHRABACHER. Um-hum.

Mr. LYONS. —are certainly very, very different from Fukushima. They are dramatically safer than Fukushima and we could certainly talk about those differences.

However, they are light water reactors. I agree with you on that point. We have in this country, we have in the world tremendous expertise on light water reactors and I don't question that there will be, I hope, a time in the future when we do move towards advanced reactors. They appear—by advanced I mean non-light water.

Mr. ROHRABACHER. Right.

Mr. LYONS. There certainly are a number of attributes that we can list that they should be able to demonstrate but I also think there is going to be more research required to get to that point. And, as I said in my opening statement, I believe that the light water reactors for the foreseeable future will be a bridge between the industry of today and an industry of tomorrow that will be able to handle and utilize the advanced reactors.

And I also—some of the other comments—there was the reference to the working group being formed within the Nuclear Energy Institute to explore advanced reactors, which I think is also very important to get industry—

Mr. ROHRABACHER. See, I don't see that as a bridge to anywhere else. What I see is this is a castle around the current establishment. I mean what we have got is not a bridge to tomorrow but a protection of the status quo. I mean your very analysis of what is going on here is we have so much expertise in the current system that it—protecting their jobs of people who now have that expertise and spent a lifetime developing it, there is something to be said to be humane to those people, but the fact is we need to have a step forward—human progress needs a nudge here, and I understand people are going to lose their jobs who don't know the new type of way of producing electricity.

It seems to me what we have, Madam Chairman, is a status quo of people who are credentialed, they have spent their lifetime learning about it, they are expert, they can be put on consulting contracts, and they don't want to change the status quo and that is why we don't ever come up with the money—we are coming up with a bridge but we never come up with the money to get across the bridge.

Chairwoman LUMMIS. I thank the gentleman.

Mr. ROHRABACHER. Thank you very much.

Chairwoman LUMMIS. By unanimous consent, the Chair now recognizes the gentleman from Georgia, Mr. Broun.

Mr. BROUN. Thank you, Madam Chairman. I appreciate the opportunity to ask a question or two, and I appreciate the opportunity to be here and thank you for serving on my Committee to and this same Committee.

But, Dr. Lyons, I am a physician from Georgia, and as you know, Georgia Power Company is building the first licensed reactor that has been approved in I guess three or four decades, and it seems to me that the Nuclear Regulatory Commission has been a huge hindrance for Georgia Power to be able to build this and it is going to cost Georgians a tremendous amount of money.

And it seems to me also that there should be a way for NRC and for DOE to have some basic schematic or preapproved plans that could be put out there for companies like Georgia Power Company or any of the Southern Company or any of the other power company in this country to be able to go ahead without having to expend so much money to get approval and have all the stoppages that have occurred over and over again.

For the name of peace, please, I beg of you try to put together some way that the power companies can build these reactors. I am a huge advocate of nuclear energy and I want to see these advanced reactors, and as we go forward with these advanced reactors, the government can be a hindrance. That is what my friend from California was talking about.

Is there any reason whatsoever that we cannot have some kind of a preapproved schematic or preapproved plans that NRC and DOE can approve and we can put these—not only the current type reactors in place but as well as the advanced reactors?

Mr. LYONS. Well, thank you for your question, Mr. Broun. I can perhaps address some aspects of that, although the majority of your question really is appropriate for the NRC.

However, as far as preapproved plans, the Vogtle plant is being built on a preapproved Part 52 design-certified AP1000.

Mr. BROUN. I understand that but over and over again NRC has caused stoppage after stoppage after stoppage, and this kind of thing is going to cost Georgians a tremendous amount of money. How can we get through this?

Mr. LYONS. Well, the idea of Part 52 is that once one has the design certification that it will be built per the way it is spelled out in the design search. To the extent that there is a departure from that, then the—Georgia Power in this case has to go back to the NRC to ask whether whatever change is being made is acceptable. I know the NRC is working on ways to streamline this process. I don't know the details since it is over in the NRC now and I am quite removed from that. But I do believe that the overall Part 52 design certification approach is the best way for the country to move forward to have certified designs where it is agreed upon up-front exactly what is going to be built, build it, and then proceed to operate it.

Mr. BROUN. Well, I promised the Chairman that I was not going to take a lot of time and I appreciate your answer.

And I beg of NRC, as well as DOE, let's make it so that we can build these nuclear plants, that we can develop the advanced reactors, and we can do so in a very cost-effective way without costing the taxpayers, as well as ratepayers, so much money.

Thank you, Madam Chairman. I yield back.

Chairwoman LUMMIS. I thank the witness for his valuable testimony.

And the Members of the Committee may have additional questions and we would ask you to respond to those in writing.

So thank you, Dr. Lyons, and you are excused.

And we will now move to our next panel.

Now, let me give you a little notice about our schedule. It looks like we are going to have a vote series coming up in 10 to 15 minutes possibly. We would like to expedite the efforts to hear what this second panel has to say so we are going to move quickly into your testimony.

This vote series is going to be long and rather than hold you here waiting for us to return, we would like to hear your testimony and then invite you back early next year so we can ask you questions about the matters to which you will be testifying.

So without further ado, it is time to introduce our second panel. Our first witness is Dr. Ashley Finan, Senior Project Manager of The Energy Innovation Project at the Clean Air Task Force. Dr. Finan manages the Advanced Nuclear Energy Project.

I would like now to ask Ms. Bonamici to introduce our second witness.

Ms. BONAMICI. Thank you very much, Chairwoman Lummis, for allowing me to participate in the hearing and I thank my colleagues on the Subcommittee who have invited Mr. McGough. Thank you for being here. Mr. Swalwell and Mr. Veasey were kind enough to invite me to introduce you.

And to Mr. McGough, thank you for your willingness to share your considerable knowledge on this issue. Mr. McGough is a 36-year veteran of the commercial nuclear industry. He has overseen the development of nuclear facilities around the globe. Now, he is helping an innovative Oregon company develop a safer approach to nuclear power.

NuScale Power is a leader in the developing field of small modular reactor, or SMR, technology. It is based in Corvallis, Oregon. My colleagues in the Oregon delegation and I have supported NuScale's efforts to secure Department of Energy funding for their design development because we see their approach as offering a safer alternative to current reactor designs.

I was pleased to see that the funding bill for Fiscal Year 2015 includes programmatic funding at the DOE that supports NuScale's design development. Following the earthquake and tsunami in Japan in 2011, my constituents expressed serious concerns about safety issues and I am very proud to have an Oregon company working to develop a safe approach to this problem.

And because I am also on the Education Committee I do want to point out that Oregon is home to the only research reactor operated primarily by undergraduate students. The research reactor at Reed College since 1968 has 40 licensed students operating it.

So, Mr. McGough, thank you for coming here from Oregon and for appearing before us today. I look forward to your testimony and I yield back. Thank you, Madam Chairwoman.

Chairwoman LUMMIS. I thank the gentlewoman.

And our third witness is Dr. Leslie Dewan, cofounder and Chief Executive Officer at Transatomic Power. Dr. Dewan was recently named one of Time Magazine's 30 people under 30 changing the world. Welcome, Dr. Dewan.

Our final witness today is Dr. Daniel Lipman, Executive Director of Policy Development at the Nuclear Energy Institute.

Now, as our witnesses should know, spoken testimony is limited to five minutes each. Hopefully, we will have time after the vote series to ask you some questions but that remains to be seen. So we are going to play it by ear.

Thank you so much, panel. Your written testimony will be included in the record of this hearing.

So I now recognize our first witness, Dr. Finan. Welcome.

TESTIMONY OF DR. ASHLEY FINAN, SENIOR PROJECT MANAGER, ENERGY INNOVATION PROJECT, CLEAN AIR TASK FORCE

Dr. FINAN. Thank you.

Chairman Lummis, Ranking Member Swalwell, and distinguished Members of this Subcommittee, thank you for holding this hearing and for giving me the opportunity to testify.

My name is Ashley Finan, Project Manager for Energy Innovation at the Clean Air Task Force. Clean Air Task Force is a nonprofit environmental organization dedicated to catalyzing the development and deployment of low emission energy technologies through research and analysis, public advocacy, leadership, and partnership with the private sector.

Climate change is an enormous challenge. To have the greatest chance of success, CATF's position is that we will need all of the low carbon energy technologies available, including nuclear power.

While nuclear technology has made big incremental improvements in the last decade and is suitable for deployment, it still faces obstacles. Advanced reactors can address those by reducing cost and construction time, enhancing safety, and better managing waste.

The United States has an exciting opportunity to continue to be a world leader in nuclear technology. We have some of the world's best innovators, a tremendous asset in the DOE and the national lab system, investors ready to invest in advance designs under the right conditions, and a regulator that is considered the global gold standard.

As with any energy technology, the development and commercialization of advanced non-light water reactors requires a suite of supportive policies from early research through demonstration and adoption. I will focus on two elements that need more attention: first, a testing facility that would enable private companies to build prototypes in a DOE-supervised environment; and second, a clear and predictable regulatory pathway for licensing advanced reactors.

Historically, the Atomic Energy Commission developed and demonstrated new reactors with full public funding on government sites. Since that level of public support was scaled back, the United States has not seen successful commercialization of a major breakthrough in nuclear reactor technology but that is not for lack of ideas. We need a new model that better incorporates private investment while taking advantage of the important role that DOE plays. A testbed facility at a DOE site would provide technology-neutral support through public-private partnership arrangements. DOE has safety oversight authority, unique capabilities, experts, and experimental facilities that could dramatically reduce the barriers, costs, and delays involved in nuclear demonstrations.

By controlling and defining many of the costs and unknowns, the testbed site would enable private investment in prototype reactors and pre-commercial projects. Not only could this unlock a great deal of private capital, it would enable U.S. innovators to move forward domestically rather than turning to foreign partners.

In addition to demonstration activities, another crucial step in commercialization is licensing with the U.S. Nuclear Regulatory Commission. The NRC's experience base is with light water technology and it has established a clear pathway for licensing a light water reactor. The process for an advanced reactor is far less established and thus introduces a level of uncertainty that can be paralyzing to private investment. Advanced reactors don't need a shortcut or less stringency but they need a well-defined, predictable process. This is another area where the model could be adjusted to enable more private and venture investment.

One such adjustment would be introducing stages of licensing. The current NRC certification process is all or nothing without interim levels of approval or acceptance. By comparison, the FDA has orderly stage dates with preclinical trials, phase 1, 2, and 3 trials; and finally, a new drug application. A drug can pass or fail at each stage and this provides a clear signal to investors that a technology is meeting or failing criteria set by the regulator.

It certainly isn't trivial to stage NRC licensing. The NRC would need resources and will, but it would provide a more workable process for investors in new technologies. In developing such a stage pathway, it would be important to collaborate closely with the innovators and investors who would use this process. There are a variety of other actions that DOE and NRC could take to develop a risk-informed and technology-neutral licensing framework that would be more applicable to advanced reactors. NRC and DOE have both taken steps in that direction but more resources and a clear mandate would ensure more timely action.

Nuclear power can play a very large role in addressing climate change, as well as other global air emissions concerns. Private investors recognize that and are ready to move forward with advanced reactors if we can modernize the commercialization model.

Thank you for this opportunity to testify. I would be pleased to respond to any questions you might have today or in the future.

[The prepared statement of Dr. Finan follows:]

Written Testimony of
Dr. Ashley Finan
Senior Project Manager for Energy Innovation
Clean Air Task Force

Before the Subcommittee on Energy
Committee on Science, Space, and Technology
U.S. House of Representatives

The Future of Nuclear Energy

December 11, 2014

Summary of Testimony

Chairman Lummis, Ranking Member Swalwell, and distinguished members of this subcommittee, thank you for holding this hearing and for giving me the opportunity to testify. My name is Ashley Finan, Project Manager for Energy Innovation at the Clean Air Task Force. Clean Air Task Force is a non-profit environmental organization dedicated to catalyzing the development and deployment of low emission energy technologies through research and analysis, public advocacy leadership, and partnership with the private sector.

Climate change is an enormous challenge. To have the greatest chance of success, CATF believes we will need all of the low-carbon energy technologies available, including nuclear power.

While nuclear technology has made big incremental improvements in the last decade and is suitable for deployment, it still faces obstacles. Advanced reactors can address those by reducing cost and construction time, enhancing safety, and better managing wastes. The US has an exciting opportunity to continue to be a world leader in nuclear technology. We have some of the world's best innovators, a tremendous asset in the DOE and the national lab system, investors ready to invest in advanced designs under the right conditions, and a regulator that is considered the global "gold standard."

As with any energy technology, the development and commercialization of advanced non-light water reactors requires a suite of supportive policies from early research through demonstration and adoption. I will focus on two elements that need more attention: First, a testing facility that would enable private companies to build prototypes in a DOE-supervised environment; and second, a clear and predictable regulatory pathway for licensing advanced reactors.

Historically, the Atomic Energy Commission developed and demonstrated new reactors with full public funding on government sites. Since that level of public support was scaled back, the US has not seen the successful commercialization of a major breakthrough in nuclear reactor technology. That is not for lack of ideas. We need a new model that better incorporates private investment, while taking advantage of the important role that DOE plays.

A test bed facility at a DOE site would provide technology-neutral support through public private partnership arrangements. DOE has safety oversight authority, unique capabilities, experts, and experimental facilities that could dramatically reduce the barriers, costs, and delays involved in nuclear demonstrations. By controlling and defining many of the costs and unknowns, the test bed site would enable private investment in prototype reactors and pre-commercial projects. Not only could this unlock a great deal of private capital, it would enable US innovators to move forward domestically, rather than turning to foreign partners.

In addition to demonstration activities, another crucial step in commercialization is licensing with the US Nuclear Regulatory Commission. The NRC's experience base is with light water technology, and it has established a clear pathway for licensing a light water reactor. The process for an advanced reactor is far less established, and thus introduces a level of uncertainty that can be paralyzing to private investment. Advanced reactors don't need a shortcut or less stringency, but they need a well-defined predictable process.

This is another area where the model could be adjusted to enable more private and venture investment. One such adjustment would be introducing stages of licensing. The current NRC certification process is "all or nothing," without interim levels of approval or acceptance. By comparison, the FDA has orderly stage-gates, starting with pre-clinical trials, Phase I, II, and III trials, and finally a new drug application. A drug can pass or fail at each stage, and this provides a clear signal to investors that a technology is meeting or failing criteria set by the regulator.

It certainly isn't trivial to "stage" NRC licensing – the NRC would need resources and will. But it would provide a more workable process for investors in new technologies. In developing such a staged pathway, it would be important to collaborate closely with the innovators and investors who would use this process.

There are a variety of other actions that DOE and NRC could take to develop a risk informed and technology neutral licensing framework that would be more applicable to advanced reactors. NRC and DOE have both taken steps in that direction, but more resources and a clear mandate would ensure more timely action.

Nuclear power could play a very large role in addressing climate change as well as other global air emissions concerns. Private investors recognize that and are ready to move forward with advanced reactors, if we can modernize the commercialization model.

Thank you for this opportunity to testify. I would be pleased to respond to any questions you might have, today or in the future.

Full Written Testimony

Chairman Lummis, Ranking Member Swalwell, and distinguished members of this subcommittee, thank you for holding this hearing and for giving me the opportunity to testify. My name is Ashley Finan, Project Manager for Energy Innovation at the Clean Air Task Force. Clean Air Task Force is a national non-profit environmental organization dedicated to catalyzing the development and global deployment of low emission energy technologies through research and analysis, public advocacy leadership, and partnership with the private sector.

Climate change is an enormous challenge. To have the greatest chance of success, CATF believes it will be necessary to deploy all of the low-carbon energy technologies available, including nuclear power.

Nuclear Energy Innovation

Working with partners, CATF helped to assemble an informal group of advanced reactor stakeholders called the Nuclear Innovation Alliance.

The Nuclear Innovation Alliance was established to develop and advocate for new federal policy initiatives and private financing to develop and commercialize advanced nuclear reactors encompassing a broad range of new and innovative designs, fuel cycles, materials, and waste characteristics. The Alliance's specific near-term objectives include adjustments that maintain rigor while making the US nuclear licensing process workable for a range of advanced reactors; public financial support for advanced reactor development and testing; and innovative private financing alliances to support advanced nuclear reactor development and commercialization. The Alliance includes environmental organizations, academic and other independent nuclear energy experts, developers of innovative nuclear reactors, and other stakeholders.

While I'm not representing the Nuclear Innovation Alliance in this testimony, the work of that group has been instrumental in developing the ideas that are described here, and I want to acknowledge the contributions of time made by all of the participants.

Climate change is a global issue, since most of the future energy demand growth and emissions will come from the developing world. The US has an obvious role in leadership, but also a crucial role in driving technology innovation so that competitive, scalable, low carbon technologies are available for global export and use.

While nuclear technology has made major incremental improvements in the last decade and is suitable for deployment, it still faces obstacles. Advanced reactors[1] can address those obstacles by reducing cost and construction time, enhancing safety, and better managing wastes. The US has an exciting opportunity to continue to be a world leader in nuclear technology. We have some of the world's best innovators, a tremendous asset in the DOE and the national lab system, investors ready to invest in advanced designs under the right conditions, and a regulator that is considered the global "gold standard."

As with any energy technology, the development and commercialization of advanced non-light water reactors requires a suite of supportive policies from early stage research through demonstration and adoption. I will focus my comments on two elements that need more attention: First, a testing facility that would enable private companies to build prototype reactors in a DOE-supervised environment; and second, a clear and predictable regulatory pathway for licensing advanced nuclear reactors.

Historically, the Atomic Energy Commission developed and demonstrated new reactor technologies with full public funding on government sites. Since that level of public support was scaled back, and essentially since the light water reactor was developed, the United States has not seen the successful commercialization of a major breakthrough in nuclear reactor technology. That is not for lack of ideas. We need a new model that better incorporates private investment, while taking advantage of the important role that DOE plays.

A Test Bed Facility for Private Investment in Reactor Prototypes

A test bed facility at a DOE site would provide technology-neutral support through public private partnership arrangements. DOE has safety oversight authority, unique capabilities, experts, and experimental facilities that could dramatically reduce the barriers, costs, and delays involved in nuclear demonstrations. By controlling and defining many of the costs and unknowns, the test bed site would enable private investment in prototype reactors and related pre-commercial projects. Not only could this unlock a great deal of private capital, it would enable US innovators to move forward domestically, rather than turning to foreign partners.

Specifically, the services and facilities that could be provided by a DOE test bed facility might include: safety oversight, water and power supply, transportation and

---

[1] In this document, the term "advanced reactors" refers to those technologies that do not use conventional water as a coolant and moderator. Small Modular LWRs occupy a commercialization stage intermediate to large LWRs and advanced non-light water reactors. They do have a few challenges in common with advanced reactors, but require minimal technological development and relatively modest regulatory adjustments. However, resolving the regulatory issues faced by small modular reactors will also assist advanced reactor developers in some ways.

security infrastructure, fuel handling, seismically characterized sites, post-irradiation testing facilities, expert consultation, etc. Which services would be provided, and at what cost, would need to be worked out among the partners. A feasibility study is needed to lay out the options for implementation, ownership, and fee structure, and to estimate the costs of building and operating such a "test bed" facility. The study should be undertaken by a consortium of the DOE, the advanced nuclear industry, potential investors, the national labs, and perhaps others, in order to ensure that all relevant perspectives are incorporated.

Advanced Reactor Licensing Developments

In addition to prototyping and demonstration activities, another crucial step in commercialization is licensing with the US Nuclear Regulatory Commission. The NRC's experience base is with Light Water Reactor (LWR) technology, and it has established a clear pathway for licensing a light water reactor. The process for licensing an advanced reactor is far less established, and thus introduces a level of uncertainty that can be paralyzing to private investment. Advanced reactors don't need a shortcut or less stringency, but they need a well-defined predictable process.

In the past, DOE has provided funding and support to help LWRs navigate the NRC licensing process, for example with the AP600 and AP1000. This same strategy might be a practical way to help advanced reactors, but it is not technology-neutral.

This is another area where the model could be adjusted to enable more private and venture investment in nuclear technology. One such adjustment would be introducing stages of licensing. The current NRC design certification process is "all or nothing," without interim levels of approval or acceptance. By comparison, the FDA has orderly stage-gates, starting with pre-clinical trials, Phase I, II, and III trials, and finally a new drug application. A drug can pass or fail at each stage, and this provides a clear signal to investors that a technology is meeting or failing criteria set by the regulator.

It certainly isn't trivial to "stage" NRC licensing – the NRC would need resources and will. But it would provide a more workable process for investors in new technologies. Over the past 25 years, the NRC developed and demonstrated a new regulatory process under 10 CFR Part 52. This "Part 52" process created separate independent approvals for the Site Permit and the Design Certification and Operating License. The stages that would help to enable advanced reactor licensing would be within the design review and operating license processes, and might be introduced into the older "Part 50" process, the Part 52 process, or a new risk-informed, technology neutral process. This is something that the NRC could pursue on its own, if it had funding to do this work,[2] or it might be something that DOE

---

[2] The NRC operates on a 90% fee recovery basis, and it isn't appropriate to use fees recovered from operating LWRs to support NRC research into advanced reactor regulation. Congress would need to appropriate funds, either to NRC under one of

could partner on, either as an extension of the NGNP licensing effort, or as an expansion of the current Advanced Reactor Licensing Initiative. In either case, in developing such a staged pathway, it would be important to collaborate closely with the innovators and investors who would use this process. The test bed site might provide a good platform for the NRC to learn about the advanced technologies and to test new licensing frameworks on an existing facility that is built and operated under DOE authority, but work to improve the licensing process should not wait for test facilities – it should begin right away.

There are a variety of other actions that DOE and NRC could take to further develop a risk informed and technology neutral licensing framework that would be more applicable to advanced reactors. NRC and DOE have both taken helpful steps in that direction,[3] but more resources and a clear mandate would ensure more timely action.

Nuclear power could play a very large role in addressing climate change as well as other global air emissions concerns. Private investors recognize that and are ready to move forward with advanced reactors, if we can modernize the commercialization model.

Thank you for this opportunity to testify. I would be pleased to respond to any questions you might have, today or in the future.

---

the exemptions to the fee-recovery requirement, or to DOE (as an NRC customer), to fund work on regulatory research. Cost estimates will need to be developed by the NRC, but we think that as little as $5-10 million annually would provide the resources to make great progress in this work.

[3] For example, through the NGNP program, the Advanced Reactor Licensing Initiative, the SMR licensing efforts, and NRC's 2012 publication: NUREG-2150 "A Proposed Risk Management Regulatory Framework."

Biography

Dr. Ashley Finan serves as senior project manager of the Energy Innovation Project at Clean Air Task Force. In that role she manages CATF's advanced nuclear energy project, coordinates studies on energy innovation policy and on advanced energy technologies. She works with the CATF team to develop and implement strategies that facilitate the commercialization of promising energy technologies.

Ashley earned her Ph.D. in Nuclear Science and Engineering at the Massachusetts Institute of Technology. Her doctoral work focused on energy innovation investment and policy optimization, both in nuclear and renewable energy technologies. She has played a key role in studies of the use of advanced nuclear energy to reduce greenhouse gas emissions in several applications, including hydrogen production, coal to liquids processes, and oil production methods. Ashley has worked as a strategy and engineering consultant, primarily on nuclear energy applications. She also contributed to an analysis of the techno-economic potential of energy efficiency improvements in the residential and commercial sectors and several related topics. Ashley holds an SB degree in Physics as well as SB and SM degrees in Nuclear Science and Engineering from MIT.

Chairwoman LUMMIS. Thank you, Dr. Finan.

The Chair now recognizes Mr. McGough.

TESTIMONY OF MR. MIKE MCGOUGH,
CHIEF COMMERCIAL OFFICER, NUSCALE POWER

Mr. MCGOUGH. Thank you. Good morning.

My name is Mike McGough and I am the Chief Commercial Officer at NuScale Power, the leading developer of American small modular reactor, or SMR, nuclear technology.

I want to thank the Committee for the opportunity to testify before you today and I want to particularly thank Representative Bonamici for her welcome and introduction. I would also like to thank Representative Veasey and Representative Weber for their continued interest and support for our work.

For 15 years our innovative company, based in Corvallis, Oregon, and majority-owned by the Fluor Corporation, has been advancing a unique SMR design that can play a significant role in our future needs for baseload carbon-free electricity generation. The NuScale design offers the safest nuclear technology available today.

[Slide]

Mr. MCGOUGH. As you see on Slide 1, we have solved one of the most vexing problems of the nuclear industry with a design approach that we call the Triple Crown of Nuclear Safety. In the event of a station blackout resulting in a complete loss of electricity comparable to what occurred at Fukushima, the NuScale Power module shuts itself down and cools for an indefinite period of time with no electricity, no operator action required, and no additional water other than an existing 8 million gallon pool. This is possible because the NuScale design eliminates many of the electrically driven pumps, motors, and valves that large reactors rely on to protect the nuclear core. Instead, our reactor is safely cooled using three simple properties of physics: convection, conduction, and gravity to drive the flow of coolant through the reactor.

[Slide]

Mr. MCGOUGH. Slide 2 presents a visual description of this natural circulation cooling process and I am happy to provide a more detailed description of this process during the question-and-answer session.

[Slide]

Mr. MCGOUGH. Our deployment characteristics are unique, and as you can see on Slide 3, the NuScale Power module is dramatically smaller than today's pressurized water reactors. It can be factory-manufactured and transported to a site via rail, truck, or barge.

Our sites are scalable. As I mentioned earlier, each site can accommodate up to 12 NuScale Power modules. Therefore, the amount of electricity at a site is scalable to between 50 and 600 megawatts based on site needs.

Continued support from Congress and the DOE is critical to our progress. Tomorrow will mark the one-year anniversary of NuScale's selection as the sole awardee for funding in round two of the DOE's Small Modular Reactor Grant Program recently authorized by Congress. The SMR program provides NuScale with the vital cost-shared funding and support of the continued design of our reactor, as well as the cost of NRC's review of our license application. NuScale may receive up to $217 million of matching funds over five years.

Of the two grant recipients under this program, we are the only developer proceeding at full speed towards near-term commercialization. Successful licensing of SMR technology depends on sustained Congressional support through continued appropriations for this program and we ask that you continue to prioritize this work.

One of the highest risk components remaining in our project is the uncertainty of the time and the process for NRC licensing. In order to meet our customer's urgent needs, we must be in a position for commercial operations in 2024. NuScale has been engaged with the NRC on pre-application review efforts since April of 2008. We expect to submit our complete application in the second half of 2016 and the NRC plan reflects a 39-month review schedule. We are waiting for the NRC to issue the NuScale design-specific review standard, which will establish the basis for our technology review.

Because of the unique technology, to ensure timely completion it is important that we have a team of NRC staff dedicated to reviewing the NuScale application. NuScale expects a robust market demand for our technology and a line of sight to our first project— projects. We are in active negotiations for our first project known as the Utah Associated Municipal Power System's carbon-free power project, which will be sited in Idaho. We expect to deliver our first project to the owner for a price of about $3 billion with subsequent plans in the range of $2.5 billion. Energy Northwest has joined this effort and the company holds first right of offer to operate the project.

The NuScale SMR is a key part of our nation's energy future. We appreciate your past support and we ask that you continue to prioritize the development of SMR nuclear technology. Thank you.

[The prepared statement of Mr. McGough follows:]

Testimony of NuScale Power before the
Committee on Science, Space, and Technology
Subcommittee on Energy of the
U.S. House of Representatives

The Future of Nuclear Energy

Testimony provided by Michael S. McGough, Chief Commercial Officer, NuScale Power

December 11, 2014

The Future of Nuclear Energy
Summary of Testimony provided by Michael S. McGough, NuScale Power
December 11, 2014

NuScale Power is the leading developer of American Small Modular Reactor (SMR) technology. For 15 years, our innovative company, based in Corvallis, Oregon and majority-owned by the Fluor Corporation, has been advancing a unique SMR design, which offers the safest nuclear technology available today. This significant advance, coupled with the deployment characteristics of our SMR design, can play a significant role in the future needs for baseload carbon-free electricity generation.

Our design is uniquely safe. We have solved one of the most vexing problems of the nuclear industry with what we call the "Triple Crown of Nuclear Safety." In the case of a station blackout event where all sources of electricity are absent, the NuScale Power Module shuts itself down and self-cools for an indefinite period of time, with no operator action required, no additional water other than an 8-million gallon pool, and no electricity. The NuScale Power Module use simple properties of physics, convection, conduction and gravity to drive the flow of coolant in the reactor. This has been demonstrated and witnessed by the U.S. Nuclear Regulatory Commission (NRC) and is protected by patents issued or pending since 2011.

Our deployment characteristics make the NuScale Power Module an option for baseload generation. The NuScale Power Module is dramatically smaller than today's pressurized water reactors and eliminates many of the electrically -driven pumps, motors and valves necessary to protect the nuclear core. It can be factory-manufactured and transported to a site via rail truck or barge.

NuScale has a robust market demand for our technology and a line of site to our first dozen projects. We are in active negotiations for the first project known as the Utah Associated Municipal Power Authority Carbon Free Power Project, which will be sited in Idaho, with possible locations including the Department of Energy's Idaho National Laboratory site. We expect to deliver our first project of 12x50 MWe NPM's in a 600MWe (gross) plant to the owner, UAMPS, for a price of approximately $3 Billion with subsequent plants in the range of $2.5 Billion. Energy Northwest has joined this effort, and holds first right of offer to operate the UAMPS project.

Tomorrow will mark the one-year anniversary of NuScale's selection as the sole awardee for funding in round two of the DOE's Small Modular Reactor program, focusing on providing cost-share grants in support of licensing expenses. NuScale may receive up to $217MM of matching funds over 5 years, and we are the only developer proceeding full speed towards near-term commercialization. Successful completion of the SMR cost-share program depends on sustained Congressional support through continued appropriations. We appreciate your past support, and we ask that you continue to prioritize this program in a tight budgetary environment.

One of the highest risk components remaining in our project is the uncertainty of the time and process for the NRC licensing efforts. In order to meet our customers' urgent needs to deliver carbon-free baseload electricity to their grids, we must be in position for commercial operations in 2024. NuScale has been engaged with the NRC on pre-application review efforts since April of 2008. We expect to submit our Design Certification Application in the second half of 2016, and the NRC plan currently reflects a 39-month review schedule. We are currently waiting for the NRC to issue the NuScale Design Specific Review Standard, which will establish the basis for our technology review. Because of the unique NuScale technology, it is important that NRC resources are dedicated to reviewing the NuScale application to ensure timely completion.

Testimony of NuScale Power before the
Committee on Science, Space, and Technology
Subcommittee on Energy of the
U.S. House of Representatives

The Future of Nuclear Energy

Written Testimony provided by Michael S. McGough, Chief Commercial Officer, NuScale Power

December 11, 2014

NuScale Power is the leading developer of American Small Modular Reactor (SMR) Technology. For 15 years, our innovative company, based in Corvallis, Oregon and majority-owned by the Fluor Corporation, has been advancing a unique SMR design which offers the safest nuclear technology available today. This significant advance, coupled with the deployment characteristics of our SMR design, can play a significant role in the future needs for baseload carbon-free electricity generation.

The genesis of our 50 MWe integral pressurized water reactor began 15 years ago with a U.S. Department of Energy (DOE) grant through the Idaho National Laboratory and included the construction of a one-third scale electrically-heated prototype test facility to validate the safety features of the plant. This prototype has been in operational testing since 2003.

Unique Safety Features

First, I will speak about the safety features of the NuScale SMR plant design. We eliminate many of the complex systems found in existing operating nuclear power plants and replace them with natural forces of physics. The result of this unique design is a nuclear plant immune to the effects of a station blackout event, like the one we observed at Fukushima.

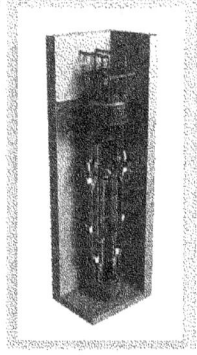

NuScale Announces Major Breakthrough in Safety
Wall Street Journal April 18, 2013

- NuScale design has achieved the "Triple Crown" for nuclear plant safety. The plant can safely shut-down and self-cool, indefinitely, with:

  - **No Operator Action**
  - **No AC or DC Power**
  - **No Additional Water**

- Safety valves align in their safest configuration on loss of all plant power.

- Details of the Alternate System Fail-safe concept were presented to the NRC in December 2012.

Nonproprietary
©2014 NuScale Power, LLC

NUSCALE POWER

As shown in this illustration, we have solved one of the most vexing problems of the nuclear industry with what we call the Triple Crown of Nuclear Safety. In the case of a station blackout event where all sources of electricity are absent, the NuScale Power Module (NPM) shuts itself down and self-cools for an indefinite period of time, with no operator action required, no additional water other than an 8-million gallon pool, and no electricity. This has been demonstrated and witnessed by the U.S. Nuclear Regulatory Commission (NRC) and is protected by patents issued or pending since 2011.

How exactly does this work?

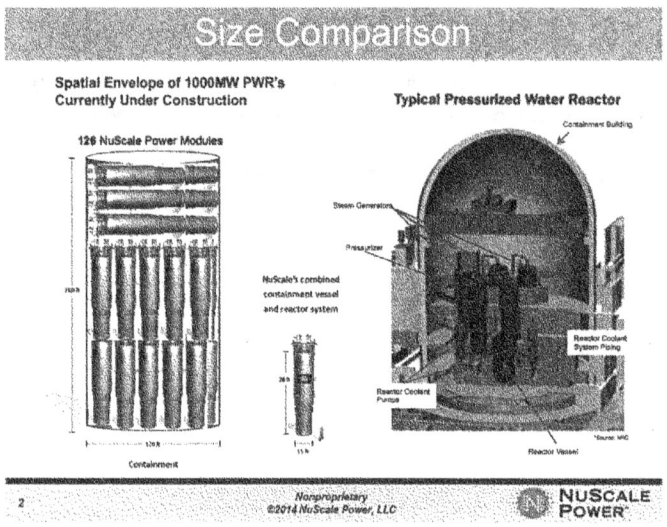

This picture illustrates the size of a 50MWe NPM compared to a typical 1000MWe-class plant operating today (on the right) and the spatial envelope of the containment buildings being constructed today in Georgia and South Carolina (on the left). The image on the right shows reactor coolant pump motors, steam generators, a reactor containment building, pressurizer and hundreds of feet of large-diameter heavy-wall reactor coolant system piping through which approximately 20 million gallons per hour of high temperature reactor coolant water flow. The center image is the NPM which is a complete integrated factory-built unit, including the containment, reactor vessel and pressurizer, all contained in one cylindrical, road-transportable vessel (Note: By design the NPM does not contain reactor coolant pumps nor large bore reactor coolant system piping).

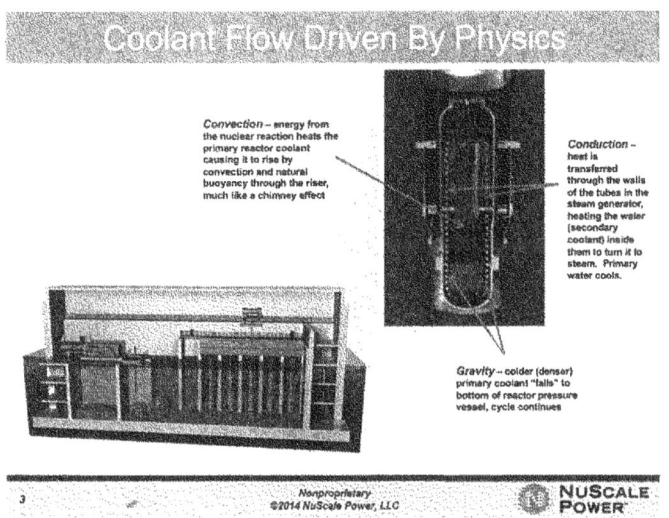

The NPM uses simple forces of physics to drive the coolant flow, as shown in this illustration. The image on the right shows the NPM which includes an outer steel vessel (containment vessel) containing an inner vessel (reactor vessel), installed underground and in an 8 million-gallon pool of water. NPMs are installed in a reactor building designed to accommodate up to 12 NPMs. The NPM operates when the heat from nuclear fission in the reactor core, represented by the red area at the bottom of the diagram, heats pressurized water causing the water to rise by buoyancy and convection through the bronze-colored tube, much like a chimney-effect. The bronze tube is surrounded by coiled tubes containing cooler water. As the hot water passes over these tubes it gives up energy by conduction through the walls of the tubes, causing the internal water to turn to steam. The steam is then directed to rotate a turbine and generate electricity. As the hot water gives up energy, it becomes cooler, thus denser, which causes it to fall by gravity to the bottom of the reactor vessel where the natural circulation flow cycle continues over and over again. The NPM eliminates many of the electrically-driven components required to protect the core in today's conventional nuclear plants.

The Role of the U.S. DOE in SMR Technology Development

In the early 2000's, Congress authorized a program known as NP-2010, to stimulate the revival of the U.S. nuclear industry with a cost-share of the private sector investments in design and licensing. This program resulted in the certification of two new nuclear plant designs, one of which is being built today in Georgia and South Carolina. Of note, the design certification testing for that design was performed under contract to the designer by NuScale founder and Chief Technology Officer, Dr. Jose Reyes, on facilities he constructed adjacent to the NuScale test facility in Corvallis, Oregon.

Congress also recently authorized a similar program for Small Modular Reactor design and licensing cost sharing. Tomorrow will mark the one-year anniversary of NuScale's selection as the sole awardee for funding in round two of DOE's SMR program, focusing on providing cost-share grants in support of licensing expenses. As such, we may receive up to $217 million of matching funds to aid in financing the approximately $1 billion necessary to complete the design and license it for construction. We have spent approximately $250 million on the project life-to-date. Successful completion of the SMR cost-share program depends on sustained Congressional support through continued appropriations. We appreciate your past support, and ask that you continue to prioritize this program in a tight budgetary environment.

A substantial portion of DOE's cost-shared funding will help pay for NRC fees required of all NRC applicants. To date, NuScale has incurred $3.3 million in NRC fees to reimburse the approximately 12,000 hours of NRC staff time utilized to date. We estimate additional NRC fees of approximately $80 million will be necessary to complete a 39 month review of our design application.

The Importance of Timely NRC Licensing Actions

In order to ensure that the NuScale design is ready to meet the urgent needs of our prospective customer base as they begin to reduce their reliance on carbon-generating assets, we will need the NRC to complete preparations for submittal of our design certification application and to conduct the design certification review in a timeframe that meets our customer's needs. We hear two consistent concerns from prospective customers and investors: 1) regulatory uncertainty, and 2) skepticism the NRC can complete their review in the 39 months the NRC has planned.

NuScale is aggressively addressing both these concerns. With respect to regulatory uncertainty we have engaged the NRC in pre-application submittal preparation since early 2008, and we have collaboratively identified about 60 remaining technical, licensing and policy issues to discuss by the end of 2015. We are currently awaiting the NRC release of the NuScale Design Specific Review Standard, which will establish the basis for our technology review. As a result of our innovative design, regulations and guidance developed for active large light water reactors do not address the many differences in our passive SMR design. In some cases, literal compliance with existing regulations would reduce the safety of our design. Timely delivery by NRC of the Design Specific Review Standard as soon as possible is critical to ensuring that we are able to submit a quality application on schedule in the second half of 2016, and meet our customers' needs. We are working closely with the NRC to ensure that this happens. NRC has signaled willingness to potentially provide us with some draft sections early, and last month we provided them a list of sections that would be most helpful.

With respect to the NRC's ability to conduct the Design Certification Application review within in 39 months, we see two critical issues to address: 1) extending the NRC's review time on critical areas by submitting portions of the design in advance; and 2) obtaining dedicated NRC staff resources to review our design. To address the first issue NuScale is submitting portions of our application early, in the form of what are known as topical reports, on about 20 aspects of the design. By submitting these reports, we effectively extend the staff's review time and reduce the scope of work that needs to be done when the remainder of the application is submitted. Regarding the second issue, the NRC reviewed the initial wave of applications using a matrix organization, where the same staff person reviewed multiple designs. While that approach is efficient and makes sense when all the designs are similar, such an approach creates more schedule risk for a unique design such as NuScale's. Therefore, we believe a dedicated NRC staff review team, whose members' sole function is to review the NuScale design, is more likely to meet the NRC's commitment to 39 months. Such a focused approach ensures that sufficient resources are identified for our review in advance of submittal and that staff can be trained on the unique safety features of our design prior to submittal of the application. We began discussions with the NRC on this approach this month.

Customers and Markets

The most important driver for our development efforts is the marketplace demand for non-carbon-generating baseload electricity. In the face of increasing carbon regulation, accelerated coal plant retirements and increased integration of intermittent renewable generation assets from wind and solar, the NPM is uniquely positioned to provide a baseload resource that is complementary to renewables and designed for load-following. And, since a NuScale plant consists of twelve individual power modules, it has the additional ability to load follow incrementally by varying the output of individual modules to match variations in intermittent generation.

The first deployment of a NuScale plant is currently designated for the Utah Associated Municipal Power System (UAMPS) for the project known as the "UAMPS Carbon Free Power Project (UAMPS CFPP)." UAMPS intends to deploy its plant for commercial operations in the 2023-24 timeframe. The plant will be located somewhere in Idaho, possibly at the DOE's Idaho National Laboratory site. Energy Northwest has joined this effort, and holds the first right of offer to operate the UAMPS project. In addition to the UAMPS CFPP, we have a line of site to our first twelve projects, including some potential additional locations at DOE sites such as Hanford, ORNL, Paducah, Portsmouth, Savannah River and others.

The NuScale plant is expected to be delivered to UAMPS for a price of approximately $3 billion, with subsequent project pricing expected to be approximately $2.5 billion. We plan to construct it on a schedule of less than 36 months from the start of safety-related construction through commissioning of the first module.

Conclusion

NuScale is proud to have developed a new technology for the deployment of a nuclear power generating asset that sets a new safety standard for the sector, is carbon-free, factory-built, incrementally deployable, and significantly less costly than large generating units. The NuScale Power Module is a disruptive technology that will change the way the world views nuclear energy, and it will play an important role in next generation deployment of baseload electricity. We are grateful for the support of the U.S. DOE and the Congress. We take those responsibilities very seriously and are singularly focused on the re-establishment of U.S. leadership in nuclear energy technology.

NuScale Announces Major Breakthrough in Safety

Wall Street Journal April 16, 2013

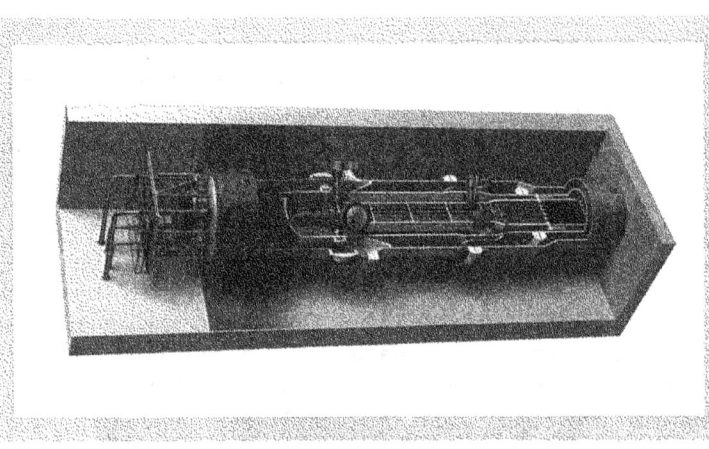

- NuScale design has achieved the "Triple Crown" for nuclear plant safety. The plant can safely shut-down and self-cool, indefinitely, with:

  - **No Operator Action**
  - **No AC or DC Power**
  - **No Additional Water**

- Safety valves align in their safest configuration on loss of all plant power.

- Details of the Alternate System Fail-safe concept were presented to the NRC in December 2012.

NUSCALE POWER™

Size Comparison

Typical Pressurized Water Reactor

Spatial Envelope of 1000MW PWR's Currently Under Construction

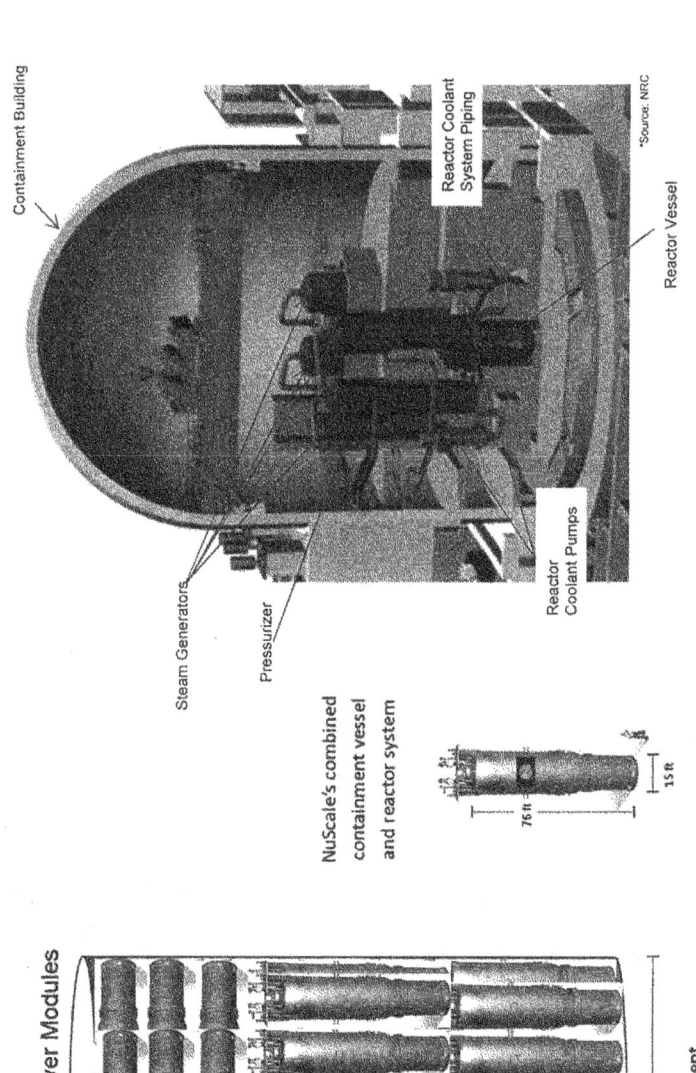

Containment Building

Steam Generators

Pressurizer

Reactor Coolant System Piping

Reactor Coolant Pumps

Reactor Vessel

*Source: NRC

126 NuScale Power Modules

NuScale's combined containment vessel and reactor system

76 ft

15 ft

120 ft

200 ft

Containment

Coolant Flow Driven By Physics

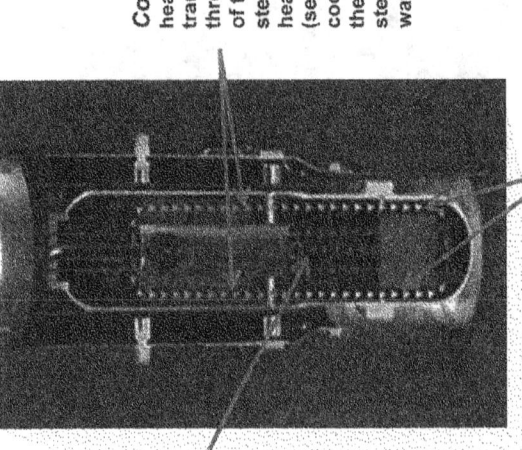

Conduction – heat is transferred through the walls of the tubes in the steam generator, heating the water (secondary coolant) inside them to turn it to steam. Primary water cools.

Gravity – colder (denser) primary coolant "falls" to bottom of reactor pressure vessel, cycle continues

Convection – energy from the nuclear reaction heats the primary reactor coolant causing it to rise by convection and natural buoyancy through the riser, much like a chimney effect

NUSCALE POWER

3

59

Biography

Michael S. McGough
Chief Commercial Officer

In his role at NuScale, Mike McGough oversees all sales, marketing, business development, proposal management, tradeshows, website, and branding.

McGough joined NuScale from UniStar Nuclear where he was Senior Vice President, Commercial Operations, with responsibilities for business development, communications, and advocacy.

McGough is a 35-year veteran of the commercial nuclear industry, working as a vendor for companies supporting nuclear plants worldwide. He has been involved in new plants at Westinghouse and UniStar, dry fuel storage as senior vice president for NAC International, low level waste management and decommissioning at Duratek and Energy Solutions. He was senior vice president then President of PCI Energy Services where he spent 11 years working mechanical projects including over 20 steam generator replacements and 10 decommissioning projects.

McGough holds a Bachelor of Science in Physical Metallurgy from Washington State University, and MBA from the Katz School of Business at the University of Pittsburgh. He is also a graduate of the Kellogg Management Institute at Northwestern University.

6650 SW Redwood Lane, Suite 210   Office 503.715.2238   Fax 503.746.6041   www.nuscalepower.com

NUSCALE POWER

Chairwoman LUMMIS. I thank the witness.

And the Chair now recognizes our next witness, Dr. Dewan.

**TESTIMONY OF DR. LESLIE DEWAN,
CO-FOUNDER AND CHIEF EXECUTIVE OFFICER,
TRANSATOMIC POWER**

Dr. DEWAN. Thank you, Chairman Lummis, Ranking Member Swalwell, and Members of the Subcommittee. I really appreciate the opportunity to be here and to talk with you all today about the future of nuclear energy and the best ways for our country to retain its superiority in nuclear technology.

I am the cofounder and CEO of Transatomic Power, a nuclear reactor design startup based in Cambridge, Massachusetts. We are developing an advanced nuclear reactor that can consume nuclear waste reducing its radioactive lifetime while generating enormous amounts of electricity.

In addition to Transatomic, there is a flourishing of other advanced nuclear reactor designs in this country that can safely produce very large amounts of carbon-free electricity with minimal waste. However, this great technology will only be useful if we can find a way to develop and commercialize it. Currently, the largest areas are the following: first of all, the lack of a clear regulatory pathway for advanced reactor development in the United States; and secondly, the lack of facilities for prototyping advanced reactor designs.

The commercial regulatory structure in the United States is currently set up only for light water reactors. The system works well for these designs but it needs to be broadened to successfully encompass advanced reactors as well. Informal estimates suggest that it would take approximately 20 years at a minimum before such a regulatory pathway for advanced reactors would be available in the United States. And furthermore, there is a great deal of uncertainty in how much regulatory approval will cost the company commercializing the design. Estimates for licensing just the prototype facility through the NRC—this is just a prototype facility—range from $200-$500 million and there are no good estimates for the cost of a commercial license for an advanced nuclear reactor.

This high cost and long timeline and furthermore the uncertainty in the estimates of the cost and timeline effectively block large-scale private investment in new nuclear reactors because no investor would want to put money into a project if they don't have a good sense of when they are going to get a return or how much it will cost at the beginning.

The current system incentivizes reactor designers to develop their first projects outside of the United States, and in fact this has already happened. Some existing nuclear reactor design companies are planning on building their first power plants overseas in Canada or China or the Philippines because they don't think it will be possible to build an advanced reactor in the United States under the current regulatory system.

A good path forward would be to move to a set of technology-agnostic guidelines based on performance criteria that would be equally applicable to all reactors. A similar set of functional guidelines, functional regulations were recently adopted in Canada and they are driving significant advanced reactor progress in that country.

Now, regulatory issues are closely tied to the ability to build prototype nuclear reactors. The great deal of uncertainty in the cost and timeline for regulating and licensing prototype nuclear facilities is a significant barrier to private investment. A clear way to solve this problem would be to establish a testbed facility ideally at a national laboratory site for building demonstration-scale advanced reactors. This solution would require clarifying the existing rules that say it is possible to build and operate demo-scale advanced reactors at national laboratory sites under the auspices of DOE without requiring an explicit license from the NRC ahead of time. NRC staffers could potentially be stationed at the site so they could observe the construction and operation of the facility. And as they do this, the NRC staffers would be building up the necessary expertise in the technology to license commercial-scale plants in the future.

Developing a better regulatory pathway for advanced nuclear reactors is vital for this country. The United States currently has the best nuclear technology in the world but I worry that this will not always be the case, especially if the most advanced reactor technology is forced to go overseas to be prototyped, licensed, and commercialized. A regulatory pathway for advanced reactors, coupled with the ability to more readily demonstrate reactor prototypes at national laboratories, will enable greater private investment in the suite of new nuclear reactor designs currently being developed and allow the United States to retain the extraordinary benefits of this new nuclear technology.

Thank you all so much. I am very, very glad to have the opportunity to testify here today and I am really looking forward to answering your questions.

[The prepared statement of Dr. Dewan follows:]

Testimony before the
U.S. House Committee on Science, Space, and Technology
Energy Subcommittee
The Future of Nuclear Energy

Dr. Leslie Dewan
Co-Founder and CEO
Transatomic Power Corporation
11 December 2014

Thank you Chairman Lummis, Ranking Member Swalwell, and members of the Committee. I greatly appreciate the opportunity to be here and talk with you today about the future of nuclear energy, the regulatory challenges for private investment and development of new reactor concepts in the United States, and the best ways for our country to retain its superiority in nuclear reactor technology. It's a pleasure to be here, and truly an honor.

I am the co-founder and CEO of Transatomic Power, a nuclear reactor design startup based in Cambridge, Massachusetts. We're developing a new type of nuclear reactor that can run entirely on nuclear waste. It consumes the waste, reducing its radioactive lifetime while generating enormous amounts of electricity.

In addition to Transatomic, there's a flourishing of other advanced nuclear reactor designs in this country – coming from industry, academia, and national laboratories – that can safely produce enormous amounts of carbon-free electricity with minimal waste.

This new nuclear technology has a great deal of potential to help the country, but only if we can find a way to develop and commercialize it. Currently, the largest barriers are the following: (1) the lack of a clear regulatory pathway for advanced reactor development in the United States, and (2) lack of facilities for prototyping advanced reactor designs. We need to develop solutions to these problems if we want to take advantage of the immense benefits of advanced reactors in the United States. I'll speak first about the regulatory pathway.

Regulatory Pathway

The commercial nuclear regulatory structure in the United States is currently set up only for light water reactors, 100 of which are operating in this country. The regulatory system works well for light water reactors, but it needs to be broadened to successfully encompass advanced reactors as well, so that the U.S. can start taking advantage of the benefits of these new designs. Right now, there is no viable pathway for bringing advanced nuclear reactor designs beyond laboratory-scale development.

Informal estimates discussed at a recent advanced reactor meeting in Washington, D.C. suggest that it would take approximately 20 years – at a minimum – before such a regulatory pathway would be available in the United States. This is a major delay. Furthermore, there's a great deal of uncertainty in how much regulatory approval will cost the company commercializing the design. Estimates for licensing just a prototype facility through the Nuclear Regulatory Commission range from $200 million to $500 million. A commercial license would cost significantly more, and there are no good estimates for what the commercial licensing cost would be for an advanced reactor.

This high cost and long timeline – and furthermore, the uncertainty in the estimates of cost and timeline – effectively blocks large-scale private investment in new nuclear reactor designs. Investors of course won't put their money into a project without good numbers for how long it will take and how much it will cost.

The current system incentivizes reactor designers to develop their first products outside of the US. In fact, this has already happened: some existing nuclear reactor design companies are planning on building their first power plants overseas – in Canada, China, or the Philippines – because they do not think it will be possible to build an advanced reactor in the US under the current regulatory system.

Talking with other advanced reactor developers in the US, we all want to be able to build our reactors this country, and license them under the NRC – NRC licensing is known as the gold standard worldwide. We'd like to find a way to adapt the system so that the NRC's high standards can be used to regulate advanced reactors as well. A good path forward would be to move to a set of technology-agnostic guidelines based on functional criteria, such as maximum radiation at the site boundary during an accident. A similar set of functional regulations were recently adopted in Canada to govern their reactor licensing processes, and is driving significant advanced reactor progress in that country. Ideally, these guidelines could be developed in a coordinated effort by the Nuclear Regulatory Commission and the Department of Energy, to effectively combine the NRC's licensing experience and the DOE's advanced reactor experience.

Test-Bed Facility

Regulatory issues are closely tied to the ability to build prototype nuclear reactors. Data from an operating prototype reactor is necessary to get a reactor licensed. Regulatory approval is necessary to build a prototype facility. Even though there are a specific subset of the NRC regulations addressing prototype facilities, there are very few data points for using these regulations for advanced nuclear reactors, and, as mentioned previously, there is a great deal of uncertainty in the cost and timeline for licensing a demonstration-scale facility. It's a chicken and egg problem that is effectively blocking new reactor development.

A clear way to solve this problem would be to establish a national test-bed facility to make it easier to build demonstration-scale advanced reactors in the US. This solution would require clarifying the existing rules that say it is possible to build and operate demo-scale advanced reactors at national laboratory sites under the auspices of DOE without requiring an explicit license from the NRC.

Under such a system, it would be possible to build a demonstration-scale reactor – not a facility that produces electric power, but simply a facility demonstrating the nuclear components of the system – at a national lab, such as the Idaho National Lab, under the auspices of DOE. NRC staffers could potentially be stationed at the site, so that they could observe the construction and operation of the facility. As they do this, the NRC staffers would be building up the necessary expertise to license commercial-scale plants in the future.

Such a plan would require only clarification of the existing regulations, and it would make a universe of difference for advanced reactor designers. It could significantly reduce the cost and timeline of licensing an advanced reactor and give greater certainty to these numbers, making it much more straightforward to raise private capital to fund them. In turn, the operating prototype facility would produce the mechanical, materials, and neutronics data necessary to license a commercial-scale facility under NRC guidelines.

Developing a better regulatory pathway for advanced nuclear reactor is vital for this country. The United States currently has the best nuclear technology in the world, but I worry that will not always be the case, especially if the most advanced reactor technology is forced to go overseas to be prototyped, licensed, and commercialized. A regulatory pathway for advanced reactors, coupled with the ability to more readily demonstrate reactor prototypes at national laboratories, will enable greater private investment in the suite of new nuclear reactor design currently being developed in this country, and allow the US to retain the extraordinary benefits of this new nuclear technology. Thank you very much. I'm very glad to have had the opportunity to testify here today, and I'm looking forward to answering your questions.

Dr. Leslie Dewan
Co-Founder and CEO
Transatomic Power Corporation

Dr. Leslie Dewan is a co-founder and the Chief Executive Officer of Transatomic Power, a nuclear reactor design company developing safe and environmentally-friendly power plants that run entirely on nuclear waste. The reactors consume the waste, reducing its radioactive lifetime while generating enormous amounts of electricity. She received her Ph.D. in nuclear engineering from MIT in 2013, with a research focus on computational nuclear materials. She also holds S.B. degrees from MIT in mechanical engineering and nuclear engineering. Before starting her Ph.D., she worked for a robotics company in Cambridge, MA, where she designed search-and-rescue robots and equipment for in-field identification of biological, chemical, and nuclear weapons. Leslie has been awarded a Department of Energy Computational Science Graduate Fellowship and an MIT Presidential Fellowship. She was named a TIME Magazine "30 People Under 30 Changing the World" in December 2013, an MIT Technology Review "Innovator Under 35" in September 2013, and a Forbes "30 Under 30" in Energy in December 2012.

Chairwoman LUMMIS. And we are looking forward to having a little bit of time to do that. So I am glad that things on the Floor are slowing down.

Our final witness is Mr. Lipman and he was nodding his head during some of the other presentations so I am looking forward to hearing his remarks.

You are recognized, Mr. Lipman. Welcome.

**TESTIMONY OF MR. DANIEL LIPMAN,
EXECUTIVE DIRECTOR, POLICY DEVELOPMENT,
NUCLEAR ENERGY INSTITUTE**

Mr. LIPMAN. Thank you, Madam Chair, and thank you, Mr. Swalwell and other Members of the Committee.

I am Dan Lipman, Executive Director at the Nuclear Energy Institute. Before joining NEI, I spent more than 31 years with Westinghouse and a period of that time included leading the new build program, the new reactor business that brought the AP1000 advanced design nuclear reactor to market.

We are keen to address the interests of the Committee. I think they are critical, particularly because they touch on three significant areas. These issues are global, these issues are long-term, and above all, they impact U.S. leadership. As for the statistics on nuclear energy is contribution in the United States, both the Chair and Dr. Lyons underlined them. I don't need to repeat them, but I will say that nuclear power plants provide a number of other key attributes, including price stability, technological diversity, and grid stability.

Nuclear fuel is not dependent on the weather or consistent fuel delivery by trucks or pipelines. During this year's polar vortex, while other electricity sources faced the challenge of crowded pipelines or even frozen fuel, the nation's nuclear power plants operated at a daily average capacity of 95 percent and no other source of electricity came close to achieving that level of reliability.

In addition to being clean, safe, and reliable, nuclear power in the United States is a tremendous demonstration of U.S. leadership. U.S. reactor designs are the basis for many of the world's nuclear power programs, yet today we face very serious competition in world markets. Major growth in nuclear energy in the near term will be outside this country, so we need to develop a program that meets this competition head-on by improving our export control processes, establishing 123 trade agreements with prospective partner countries, reauthorization of the Export-Import Bank, and importantly, as you have heard today, continuing to develop the safest and most advanced nuclear technologies here.

And while we need to compete abroad, leadership means we need to be building and developing more nuclear at home. Maintaining nuclear energy's share requires the equivalent of 12 new nuclear power plants by 2025, and if today's nuclear power plants retire at 60 years of operation, we will need 20 plants by 2030 and 45 by 2035, so subsequent license renewal is critical.

It is a strategic imperative to deploy small modular reactors in the early to the mid-2020s followed by more advanced generation designs in the 2030s and beyond. Small modular reactors allow capacity additions at smaller increments and advanced reactors will likely have an even higher level of inherent safety and may be able to serve a vital role in management of spent fuel from today's light water reactors.

Commercialization of advanced nuclear reactors will best be achieved through an appropriate program that identifies these technologies, facilitates their deployment, and as you have heard, the most significant challenges facing both SMRs and Generation IV reactors are financing and licensing. The time, the uncertainty, and the cost required to design, license, and build new reactors is daunting.

You heard from Secretary Lyons that at NEI we are establishing, similar to our SMR working group, an advanced reactor working group chaired by the CEO of Southern Nuclear Operating Company to develop an industry vision of a long-term sustainable program that will support the development and commercialization of advanced reactors. We must establish a portfolio of technologies necessary to provide clean, reliable baseload electricity for the 2030s and beyond. federal and state governments and industry must address in the balance of this decade—so in the next five years—the financing and regulatory challenges facing these advanced nuclear technologies.

Both SMRs and Gen IV reactors need to have their barriers deployment and eventually, as you have heard, for overseas markets. We need innovation, creative approaches to ensure the availability of capital and regulatory certainty and closure. Business as usual will not get the job done.

Thank you.

[The prepared statement of Mr. Lipman follows:]

STATEMENT FOR THE RECORD
by
Daniel S. Lipman
Executive Director, Policy Development and Supplier Programs
Nuclear Energy Institute
to the
Committee on Science, Space and Technology
Subcommittee on Energy
U.S. House of Representatives

December 11, 2014

The Nuclear Energy Institute thanks the House Science Committee for its interest in nuclear energy and in addressing the policies that can facilitate deployment of advanced reactors to meet national energy needs and reduce carbon emissions.

My name is Daniel Lipman. I am Executive Director for Policy Development and Supplier Programs at the Nuclear Energy Institute (NEI). NEI is responsible for establishing unified nuclear industry policy on regulatory, financial, technical and legislative issues affecting the industry. NEI members include all companies licensed to operate commercial nuclear power plants in the United States, nuclear plant designers, major architect/engineering firms, fuel cycle companies, and other organizations and individuals involved in the nuclear energy industry. Before joining NEI, I spent 31 years with Westinghouse Electric Corporation, a period that included leadership of the company's program to bring the AP1000 advanced-design nuclear reactor to market. Four AP1000 reactors are now under construction in China, and four are being built in the United States.

My testimony will cover five major areas:

1. The current status of the U.S. nuclear power industry and the value proposition for nuclear energy;
2. The potential domestic market and our conviction that nuclear energy will be a major part of the future U.S. supply portfolio;
3. The global nuclear market and U.S. influence;
4. The potential for Small Modular Reactors and Generation IV designs, and
5. The absolute necessity of effective government-industry cooperation to address financing and regulatory challenges

Before we explore the subject of this hearing – the future of nuclear energy – it's appropriate that we discuss the importance of electricity and review how the nuclear industry arrived at where it is today.

The International Energy Agency, in its World Energy Outlook, emphasizes the importance of energy production:

> "Energy is a critical enabler. Every advanced economy has required secure access to modern sources of energy to underpin its development and growing prosperity. In developing countries, access to affordable and reliable energy services is fundamental to reducing poverty and

improving health, increasing productivity, enhancing competitiveness and promoting economic growth. This is because it is essential for the provision of clean water, sanitation and healthcare, and provides great benefits to development through the provision of reliable and efficient lighting, heating, cooking, mechanical power, transport and telecommunication services."

The World Energy Outlook 2014 estimates that 1.3 billion people worldwide live without access to electricity. That's about 1 out of every 5 people in the world and larger than the combined population of North and South America. To reduce poverty and to raise the standard of living for all, the world needs to produce more electricity. There is no single technology that can accomplish this task. Nuclear power, which doesn't produce pollutants and is reliable source of baseload electricity, must be a significant part of the electricity mix worldwide.

The history of nuclear power in this country is a remarkable story of leadership, innovation and excellence in reactor design and operation that continues to this day. U.S leadership introduced this technology to the world and is responsible, directly or indirectly, for most of the nuclear programs in the world.

U.S. reactor designs – both pressurized water reactors and boiling water reactors – are the basis for the French nuclear program, the Japanese nuclear program, the South Korean and Chinese nuclear programs. The British, after building a commercial program based on gas-cooled reactors, turned to American-style PWR technology in the 1980s. This tradition of technology leadership continues today, with deployment in China and the United States of U.S. designed AP1000 advanced-light water reactors that incorporate passive safety features – the AP1000 and the ESBWR are the most advanced designs currently available from any nation or any vendor. This tradition of leadership continues with the design and development of small modular reactors (SMRs) and even more advanced reactors, Generation IV (Gen IV) reactors.

In addition to our technological leadership, the United States has the best operating experience in the world, with the U.S. nuclear fleet consistently recording average capacity factors in the 90-percent range, year in and year out, since the late 1990s. Supporting this success is a unique infrastructure – the various programs managed by the Institute for Nuclear Power Operations (INPO) – designed to maintain excellence in operations. The government's review of the factors that led to the Deepwater Horizon disaster and oil spill in the Gulf of Mexico in 2010 pointed to INPO as the best example of an industry organization designed to establish and maintain high standards of operating excellence.

Nuclear energy in the United States is also one of the few energy technologies – if not the only technology – that fully internalizes its costs, including decommissioning and waste management.

Finally, let's remember that the Nuclear Regulatory Commission is widely regarded as the "gold standard" of regulatory agencies worldwide, and that a design certification from the NRC is considered an unimpeachable seal of approval.

This brings us to today. And today, we face serious competition in world markets from Russia, China, South Korea and France. Many of these countries are competing against us with the same technology that we transferred to them in years past. We must adapt to that competition and meet it head-on, recognizing that the major growth in nuclear energy in the near-term will be overseas, and that we must improve our export control process and provide competitive financing, in addition to the best technology.

We must also adapt to fiscal reality, recognizing that federal government dollars are limited. In the early years of nuclear power in America, the Atomic Energy Commission financed reactor development and demonstration directly. As the technology matured, development and demonstration evolved into a cost-shared government-industry effort, like the hugely successful Nuclear Power 2010 program, which gave us the AP1000 and ESBWR advanced reactor designs. Looking forward, we must apply the same innovation to financial engineering, regulatory development, and licensing that we do to reactor engineering, and develop innovative techniques to bring new technologies to market.

We can meet the challenges of the 21st century. We have the tools. We have the resources. We have the technical edge, and an economic system that encourages innovation. And we must succeed, because we cannot afford to cede U.S. leadership in commercial nuclear technology to countries like Russia and China.

I. Current Status of the U.S. Nuclear Power Industry and the Value Proposition for Nuclear Energy

In 2013, nuclear energy produced 19 percent of U.S. electricity supply (789 billion kilowatt-hours). The industry's 2013 average capacity factor was 90.9 percent, compared to 86.4 percent in 2012. This is the highest capacity factor of any source of electric power. The U.S. nuclear energy industry's top priority is, and always will be, the safe and reliable operation of our plants. Safe, reliable operation drives public and political confidence in the industry, and America's nuclear plants continue to sustain high levels of safety and performance.

NEI believes that America's nuclear energy assets provide a uniquely valuable set of attributes:

- Nuclear power plants produce large quantities of electricity around the clock, safely and reliably, when needed. They operate whether or not the wind is blowing and the sun is shining, whether or not fuel arrives daily, weekly, or even monthly by truck, barge, rail or pipeline.
- Nuclear plants provide price stability to the grid.
- They provide "reactive power" – essential to controlling voltage and frequency and operating the grid.
- Nuclear power plants have portfolio value, contributing to the fuel and technology diversity that is one of the bedrock characteristics of a reliable, resilient electric sector.
- Finally, nuclear power plants provide clean air compliance value. In any system that limits emissions – of the Clean Air Act "criteria" pollutants or carbon dioxide – the emissions avoided by nuclear energy reduce the compliance burden that would otherwise fall on emitting generating capacity.

Other sources of electricity have some of these attributes. None of the other sources has them all.

Nuclear plants are also critical to the reliability of the electric grid because they operate continuously and generally independently of weather conditions. For example, during the "Polar Vortex" event, which occurred during the week of January 6, 2014, the nation's nuclear power plants operated at daily average capacity factors of over 95 percent. No other source of electricity approached that level of reliability. In fact, approximately 25 percent of the generating capacity in the PJM Interconnection and 20 percent of the capacity dispatched by the Midcontinent System Operator (MISO) was forced out of service by the severe cold weather (generally because power plants could not obtain fuel or because fuel-handling equipment froze).

Nuclear energy's high reliability will become increasingly important as the nation's electricity system becomes less reliant on coal, and more reliant on gas-fired generating capacity and on renewable technologies that are intermittent and weather-dependent.

Nuclear energy is also America's largest source of low-carbon electricity. In 2013, nuclear energy accounted for 63 percent of America's carbon-free electricity, and prevented 589 million metric tons of CO_2 emissions – three times more carbon-free electricity than hydropower and nearly five times more than wind energy. For perspective, one gigawatt of nuclear generating capacity (out of the 100 GW operating) would avoid more carbon than all U.S. solar energy capacity in 2013 (4,500 megawatts at 17-percent capacity factor). The amount of CO_2 emissions avoided by nuclear energy facilities is equal to the CO_2 emissions from 113 million passenger cars – more than all the passenger cars in the United States.

America's 100 reactors are also a significant Clean Air Act compliance tool. They avoid approximately one million tons of sulfur dioxide and half-a-million tons of nitrogen oxide emissions annually, according to NEI calculations using protocols developed by the Energy Information Administration. Without nuclear energy, U.S. electric sector emissions of CO_2, SO_2 and NO_x would be approximately 25-30 percent higher every year.

II.   The Domestic Market: Nuclear Energy A Major Part of the Future U.S. Supply Portfolio

Even at less-than-one-percent annual growth in electricity demand, the Energy Information Administration (EIA) forecasts a need for 350 gigawatts of new electric capacity by 2040 (28-percent growth) in the United States. To satisfy this demand at lowest possible cost without compromising the nation's environmental goals, the U.S. electric power industry must have a portfolio of electricity generating technologies, particularly low-carbon technologies.

Unfortunately, trends are moving in the other direction. America's electric generating technology options are narrowing dramatically:

- Coal-fired generating capacity is declining in the face of increasing environmental restrictions, including the likelihood of controls on carbon, and uncertainty over the commercial feasibility of carbon capture and sequestration. The U.S. has about 300,000 MW of coal-fired capacity, and the consensus is that about 60,000 MW of that will shut down by 2020 because of escalating environmental requirements.[1] In addition, the pipeline of coal-fired projects under development is all but empty.

---

[1] This is the estimated coal-fired capacity likely to be shut down due to existing regulation of so-called "criteria pollutants" (e.g., SO_2, NOx, fine particulates, mercury, air toxics). The Environmental Protection Agency's proposed regulation to reduce carbon emissions from existing power plants would lead to an additional 40,000-45,000 MW of coal-fired retirements, by most estimates.

- Natural gas-fired generating capacity is growing dramatically. Since 1995, the United States has built approximately 342,000 megawatts of gas-fired capacity, approximately 75 percent of all capacity additions. Coal and nuclear, the two sources of electricity that can produce electricity around-the-clock at stable prices with virtually no price volatility, represent a scant six percent of the generating capacity added. Clearly, the United States should not continue to build *only* gas-fired generating capacity.

- Renewables will play an increasingly large role but, as intermittent sources, cannot displace the need for baseload generating capacity, absent dramatic advances in energy storage.

In this environment, the United States must have as many generating options as possible. A continuing, growing contribution from nuclear energy is essential to produce the baseload electricity that will be needed at stable prices, and to sustain reductions in emissions of carbon and other criteria pollutants. Small modular reactors and advanced reactors are an essential part of the generating portfolio of the future.

The electric power industry is approaching a period – the decade starting in 2020 – when it must consider how to replace the nuclear generating capacity that will reach the end of its 60-year licensed life starting in approximately 2030. (Just over 31,000 MW of nuclear generating capacity, of the approximately 100,000 MW operating in the United States, reaches the 60-year point by 2035.) The capital cost to replace that capacity with new nuclear generating capacity would be many billions of dollars. Some of this capacity will likely seek a second license renewal to operate past 60 years, but some will not. The regulatory framework for second license renewal has not yet been firmly established. Additional capital investment will almost certainly be required to operate past 60 years and, in some cases, market conditions or other factors may not justify that capital investment.

Figures 1 and 2 show the nuclear generating capacity required under two potential scenarios: (1) if all today's reactors operate to 60 years, then retire; and (2) if all today's reactors operate to 80 years. These calculations start with the projections in EIA's 2014 *Annual Energy Outlook*, which assumes 0.7-percent annual growth in electricity demand. Some observations:

Figure 1

New Nuclear Generating Capacity Needed If All Reactors Retire After 60 Years of Operation

Year	Total Electric Generation (bkWh)	Nuclear Capacity (GW)	Nuclear Generation (bkWh)	Nuclear Fuel Share	New Generation Needed to Meet Fuel Share (GW)	
					20%	25%
2025	4,622.3	104.0	820.0	17.7%	13.2	42.6
2030	4,815.1	100.0	788.0	16.4%	22.2	52.7
2035	5,004.3	72.4	570.4	11.4%	54.6	86.3
2040	5,219.7	57.5	453.2	8.7%	74.9	108.0

Data Source: *Energy Information Administration, Annual Energy Outlook 2014*

- Simply because of load growth, maintaining nuclear energy at 20 percent of U.S. electricity supply would require 13.2 gigawatts (GW) of new nuclear capacity by 2025 (in addition to the Watts Bar 2, Vogtle 3 and 4, and Summer 2 and 3 reactors already under construction).

- If today's nuclear plants retire at 60 years of operation, 22 GW of new nuclear generating capacity would be needed by 2030, and 55 GW by 2035, to maintain a 20 percent fuel share.

- If today's reactors operate to 80 years, 18 GW of new nuclear capacity would be needed by 2030, and 23 GW by 2035, to maintain a 20 percent share of U.S. electricity supply.

- Much larger amounts of new nuclear generating capacity would be needed to drive nuclear energy to 25 percent of U.S. electricity supply.

Figure 2

New Nuclear Generating Capacity Needed If All Reactors Retire After 80 Years of Operation

Year	Total Electric Generation (bkWh)	Nuclear Capacity GW	Nuclear Generation (bkWh)	Nuclear Fuel Share	New Generation Needed to Meet Fuel Share (GW)	
					20%	25%
2025	4,622.3	104.0	820.0	17.7%	13.2	42.6
2030	4,815.1	104.0	820.0	17.0%	18.1	48.7
2035	5,004.3	104.0	820.0	16.4%	22.9	54.7
2040	5,219.7	104.0	820.0	15.7%	28.4	61.5

Data Source: Energy Information Administration, Annual Energy Outlook 2014

Although dates like 2030 and 2040 seem like a far-distant future, in the world of electric power planning, they are not. Given the lead times necessary to license and build new nuclear generating capacity, planning for, and construction of, this capacity must begin in the early 2020s. That leaves the balance of this decade – a relatively short five years – to ensure that the necessary technologies are ready for deployment, and to put in place the policy conditions necessary to enable that deployment.

In addition, it is clear from the simple arithmetic in Figures 1 and 2 that the United States has several major nuclear-related imperatives.

- First, a workable regulatory framework for subsequent license renewal is essential.

- Second, it is essential to reduce the time to market for certified designs. Maintaining a high level of standardization when constructing the large light water reactors (the AP1000 and the ESBWR) already certified by the NRC is essential. With standardization and some reforms to the Part 52 licensing process (based on lessons learned during the licensing and construction of the new Vogtle and Summer units in Georgia and South Carolina), it should be possible to reduce time to market to from 10 years or so to seven years. Absent rigorous adherence to standardization, it will be more difficult to obtain the combined construction and operating license and build these reactors to cost and schedule in the relatively large numbers likely to be required.

- Third, advanced reactors – starting with small modular reactors (SMRs) in the early- to mid-2020s, followed by more advanced Generation IV designs in the 2030s and 2040s – are a strategic imperative. The nuclear industry will need as many technology options as possible. For example, because SMRs allow capacity additions in smaller increments, they may be particularly well-suited to regions of the country with low growth in electricity demand. And they may be the only way – or certainly the easiest way – to finance new nuclear capacity in competitive markets, because they do not require a company or companies to undertake a gigawatt-scale project, a $7-8 billion financing, in a single step.

III.   The Global Nuclear Market and U.S. Influence

Today, there are 72 new nuclear power stations under construction worldwide, of which five are under construction in the United States. An additional 172 are in the licensing and advanced planning stages and virtually all of these plants will be built abroad where the demand for reliable, affordable and clean baseload electricity is growing. Electricity from nuclear energy will help developing economies expand and lift hundreds of millions from poverty while having a minimal impact on the environment. For developed economies, nuclear energy is widely recognized as a reliable source of generation that provides significant electricity supplies without emitting greenhouse gases during operation. But with this growing nuclear market comes growing competition from other nuclear supplier nations, which can now provide a full range of products and services.

With the world's largest civilian nuclear energy program, the U.S. industry is recognized for reliability, safety and operational excellence. U.S. firms are making major investments in technology development to continue their tradition of innovation. These investments include development of small modular reactors, advanced technologies for uranium enrichment, advanced reactor designs with improved safety features and advanced manufacturing techniques to improve quality and reduce costs. Coupled with the globally recognized "gold-standard" regulator, the U.S. Nuclear Regulatory Commission, many nations place a high value on cooperation with the U.S. as they develop or expand their civilian nuclear energy programs.

Over the past two decades, new supplier nations have entered the growing global nuclear market, and multi-national partnerships and consortia have been formed to develop nuclear energy facilities. According to a 2010 GAO report, "while the value of U.S. exports of nuclear reactors, major components and minor components have increased, the U.S. share of global exports declined slightly" from 1994 to 2008.[2] Over the same period, the U.S. share in the fuel market declined sharply from one-third to one-tenth of the market.

The growth of nuclear suppliers overseas has increased competition for U.S. firms. International competitors often began as suppliers to their domestic markets and over time expanded their offerings to the global market. For example, France's AREVA and Russia's Rosatom have steadily increased their presence in the global market. Although 12 of the reactors under construction today are U.S. designs, four are French and 16 are Russian.[3] One of the newest entrants in the global nuclear market is the Republic of Korea. In December 2009, Emirates Nuclear Energy Corporation awarded a multi-billion dollar contract to a Korea Electric Power Corporation-led consortium to build the first two nuclear power plants in the United Arab Emirates (UAE). In addition, there has been an expansion of indigenous technologies developed for domestic markets. For example, 20 of the 72 nuclear plants under construction globally are Chinese reactors being built in China.[4]

---

[2] "Global Nuclear Commerce: Governmentwide Strategy Could Help Increase Commercial Benefits From U.S. Nuclear Cooperation Agreements with Other Countries", United States Government Accountability Office Report to the Committee on Foreign Affairs, House of Representatives, November 2010.
[3] International Atomic Energy Agency, 2014.
[4] Ibid.

The international market for nuclear power is growing and the U.S. industry has the opportunity to supply a significant portion of this demand with innovative technologies supported by continued investment in research and development.

IV.   Potential for Small Modular Reactors and Generation IV Designs

In the electric power business, where technology development, demonstration and deployment is a decades-long exercise, the time to develop and demonstrate new technology is years before it is needed – not when it is needed. As utilities plan for the future they must look beyond the short term and continue to adapt to the changing landscape. Over the next 30 years, a significant amount of existing generating capacity will be retired. Decisions on which technologies will replace these retirements will be made within the next 10-20 years. In the short- to medium-term, light water reactors – large gigawatt-class reactors and SMRs – will remain the dominant and most economic means of electricity production from nuclear energy.

The potential for small modular reactors and advanced reactors is enormous:

- Because they can be built in sequence, SMRs allow generating companies to match construction of new capacity with electricity load growth – particularly important in parts of the country where load growth may have slowed permanently.

- SMRs also provide financing flexibility: The capital investment can be staged as modules are constructed. This could be particularly important for smaller companies – rural electric cooperatives or municipal power agencies, for example – that cannot afford the $6 billion - $7 billion up-front financing associated with a 1,000-megawatt reactor.

- In the U.S., SMRs could be used to replace older fossil-fueled generation that will not meet new EPA clean air requirements. SMRs also provide a clean generation option for municipal and rural co-op utilities whose portfolios are dominated by older, small fossil generation facilities.

- Small reactors have enormous potential for overseas markets, particularly in countries that are developing a nuclear energy industry for the first time.

- Advanced reactors can expand the slate of products provided by a nuclear power plant – to include process heat, for example – or serve a vital role in management of the spent nuclear fuel from today's light water reactors, thus minimizing the volume of high-level waste requiring permanent disposal.

Advanced (Generation IV) nuclear reactors hold the promise of inherently safe, emission-free, low-cost energy. Advanced reactors are generally understood to be fission reactor designs that represent significant advances from light-water reactor (LWR) technologies, including advanced LWRs, in terms of resource utilization, level of inherent safety, and substantially higher operating temperatures. In addition, advanced reactors have the potential to provide services beyond simply electricity generation.

If utilities are to consider advanced reactors in their future decision-making, significant progress toward commercialization is necessary. A focused, aggressive, coordinated effort by government and industry is necessary to ensure that SMRs and Gen IV reactors are ready for deployment in the United States and overseas as quickly as possible.

Commercialization of advanced nuclear reactors will best be achieved through a heavily industry-influenced research, development and demonstration (RD&D) program. An appropriate RD&D program must be able to identify the most promising technologies, and incorporate a decision-making process to facilitate down-selection, demonstration and deployment. Such an approach will ensure that real options are available when needed and at the scale needed to meet national and global electricity requirements. To focus the industry's perspectives, the Nuclear Energy Institute is establishing an Advanced Reactor Working Group, chaired by Stephen Kuczynski, Chairman, President and CEO of Southern Nuclear Operating Company. The working group, fashioned after NEI's Small Modular Reactor Working Group, will be charged with developing an industry vision of a long-term sustainable program that will support the development and commercialization of advanced reactors, ultimately supporting the commercial availability of advanced Gen IV reactors in the 2035 to 2040 timeframe.

V. Government-Industry Cooperation is Essential to Address Financing, Regulatory Challenges

The most significant challenges facing development, demonstration and deployment of SMRs and Gen IV reactors are financing and licensing. The time, uncertainty and cost required to design, license and build a new reactor design is daunting. The United States must develop creative approaches to lower barriers to entry – including industry-government cost-shared programs, investment incentives and innovative approaches to financing design, demonstration and initial deployment, and a more efficient regulatory framework for licensing of advanced reactors.

Financing Challenges. For SMRs, reactor designers are making significant progress in developing SMR designs. The Nuclear Regulatory Commission (NRC) is prepared to license these designs, supported by a cost-shared DOE licensing support program and industry work on generic SMR licensing issues. Customer interest, from the companies that will finance and operate these plants, is strong – notably strong given the economic and financial stress under which the U.S. electric sector is laboring, and the fact that first deployment is not expected until the early 2020s, at best.

Based on growing industry interest in SMRs, the Department of Energy (DOE) issued a Funding Opportunity Announcement (FOA) for a cost-shared industry partnership program in March 2012. The goal of the SMR Licensing Technical Support (LTS) program is to accelerate commercial deployment of SMR technologies. The LTS program is funded on a 50-50 cost-shared basis by DOE and industry participants, with U.S. government support currently capped at $452 million over five years.

This program seeks to replicate the success of the Department of Energy's Nuclear Power 2010 program, which provided development funding for large advanced-design reactor technology, resulting in the design and certification of the ESBWR and AP1000 reactors. Eight AP1000 reactors are now under construction in the United States and China.

Simple business reality dictates that cost-shared programs, with a substantial federal contribution, are essential. In the case of SMRs, the cost-shared government-industry program is necessary because plant designers will not see revenue or positive cash flow for approximately 10 years – longer than most companies can tolerate. Companies like Babcock &Wilcox (B&W), NuScale and others are prepared to absorb a significant share of the technology design and development costs, but the federal government must also play a significant role – particularly given the enormous promise of SMR technology. B&W's Generation mPower joint venture, for example, has already invested $400 million in developing its mPower design; NuScale, approximately $200 million in its design. These are significant sums of money, which will not generate any return for approximately a decade.

Beyond the design and engineering work necessary for design certification, substantial sums of money are required for critical activities essential for deployment. Completion of detailed design beyond the basic design phase is necessary. This will allow the development of specification-level documentation and a detailed design for construction and for procurement.

Current estimates are that the cost to complete enough design and engineering to obtain an NRC Design Certification and a Combined Construction/Operating License (COL) is expected to be approximately $750 million to $1 billion per design. More detailed first-of-a-kind (FOAK) design and engineering – required to develop credible cost and schedule estimates, a prerequisite to utility planning and investment decisions – is expected to add approximately $500 million to the cost of obtaining NRC approvals. Financing this development is a significant challenge.

NEI believes there are approaches that could help address this challenge. In its comments to the Department of Energy on the Quadrennial Energy Review on this issue, NEI recommended that the government consider innovative approaches to close the funding gap, including use of the Title XVII loan guarantee program. For example, the design, engineering and licensing costs could be folded into the cost of the first project(s) and financed through the loan guarantee program. Since the loan guarantee program can cover up to 80 percent of total project cost, this approach could allow project developers and generating companies to use debt (which is always lower cost than equity) to finance a substantial portion of the design, engineering and licensing costs, and to repay those costs over the long period of time typical of debt maturities. This is just one example of an innovative financing technique that, in NEI's view, deserve further exploration.

Regulatory Challenges. Regulatory uncertainty and the cost of obtaining the necessary regulatory approvals are also a challenge. This is not so much because NRC regulations are unnecessarily burdensome, but because the requirements for advanced reactors are not clear, largely because NRC regulations are based on existing large light water reactor technology. NEI believes that the NRC process for licensing advanced reactors could be more efficient – producing the same level of safety, but less uncertainty, time and cost to go from application to license.

In a 2012 report to Congress on the status of its Advanced Reactor licensing activities, NRC reported that its advanced reactor program is focused on preparing the agency for reviews of applications related to the design, construction, and operation of advanced reactors. These efforts include the following major areas:

- Identify and resolve significant policy, technical, and licensing issues;
- Develop the regulatory framework to support efficient and timely licensing reviews;

- Engage in research focused on key areas to support licensing reviews, and
- Engage reactor designers, potential applicants, industry, and DOE in meaningful pre-application interactions and coordinate with internal and external stakeholders.

NRC has demonstrated a willingness to work with stakeholders early, before applications are filed, to resolve key policy and technical issues for advanced reactors. As a result, most of the important issues are well-defined and have been widely-discussed for, in some cases, over two decades. Even so, potential licensees face a "chicken and egg" conundrum that must be resolved – i.e., necessary changes to NRC policies and technical requirements for advanced reactors wait for "real" projects because NRC is reluctant to commit to those changes in the absence of a "real" project, but projects need clarity and resolution on policies and technical requirements before they can become "real."

VI.   Conclusion

For decades, nuclear and coal-based technologies have been the bedrock of the U.S. electric supply system. The coal-based options are narrowing, which creates a compelling need to ensure that the nation has available a robust, diverse suite of nuclear technologies. Failure to do so would condemn the nation to larger and larger dependence on one fuel – natural gas – for electricity production.

As America's existing generating capacity, including some portion of its nuclear generating capacity, approaches the end of its useful life, the nation must take steps to establish the portfolio of technologies necessary to produce clean, reliable baseload electricity for the 2030s and beyond. To be operational in the 2030s, this generating capacity must be under construction in the 2020s. To be under construction in the 2020s, federal and state governments and industry must address – in the balance of this decade – the financing and regulatory challenges facing these advanced nuclear generating technologies, both SMRs and Gen IV reactors.

Tackling these challenges successfully will require innovative, creative approaches to ensure availability of capital and the regulatory certainty and closure required. Business as usual will not get the job done.

■

NUCLEAR ENERGY INSTITUTE

Daniel S. Lipman
Executive Director, Policy Development and Supplier Programs

Dan Lipman joined the Nuclear Energy Institute in January 2014. He had worked there in 2012-2013 as a loaned-executive from Westinghouse Electric Company. He is responsible for nuclear exports, international trade, relations with supplier member companies, state regulatory relationships, fuel cycle policy and programs, nonproliferation, and policy analysis.

Before joining NEI, Mr. Lipman was senior vice president of operations support for Westinghouse Electric Company, responsible for the global supply chain, quality assurance and continuous improvement, information technology, corporate strategy, risk management, environmental health and safety, sustainability and leadership over a number of key commercial processes and transformational activities. He assumed this position in September 2009.

From 2005-2009, Mr. Lipman served as senior vice president, nuclear power plants, responsible for managing Westinghouse's global deployment of new, nuclear power plants. During this time, he played a pivotal role in securing landmark AP1000 contracts in China and the United States, establishing Westinghouse as the early leader of the nuclear renaissance. He had executive responsibility for AP1000 design and licensing, marketing, contract negotiations, equipment supply and project management.

Prior to 2005, Mr. Lipman was vice president, customer relations and sales, North America, with responsibility for strategic relationship management, commercial affairs, sales and alliances for nuclear operating plant customers. Under his leadership, U.S. performance for Westinghouse reached more than $1 billion of new orders for nuclear fuel and services in the U.S. alone.

Mr. Lipman also served as president of Westinghouse Asia, with regional duties for China, South Korea and Taiwan. Living in Beijing for four years, he was responsible for relationship management and sales of Westinghouse nuclear engineering, equipment and services to operating nuclear plants, plants under construction and new plants in those markets. The implementation of joint ventures, technology transfer and partnerships with local organizations were important aspects of this activity, as was establishing and developing host country government and embassy relationships.

Mr. Lipman's earlier experience involved numerous project management and customer interface assignments. He spent approximately seven years in field positions at customer plant sites in Korea, Georgia and Texas supporting construction, start-up, operations, and outage management. At headquarters, he furnished project management and technical support to key customer accounts for Westinghouse products and services at operating plants.

Other assignments have included sales support roles for Combustion Engineering plants and negotiating and implementing technology transfer agreements with European companies.

Before joining Westinghouse, Mr. Lipman was a research associate at a Washington, D.C., consulting firm for nuclear nonproliferation, reprocessing, waste management and fuel-cycle policy issues. He also interned at the International Atomic Energy Agency in Vienna.

■

Chairwoman LUMMIS. I thank the witnesses and we will now open our round of questions. The Chair recognizes the gentleman from Texas, Mr. Weber.

Mr. WEBER. Thank you, Madam Chair. Dr.—is it Dewan? Is that how you say that?

Dr. DEWAN. Yes, it is Dewan. Thank you.

Mr. WEBER. Okay. Transatomic has proposed a molten salt reactor, a non-light water design that will run on nuclear waste and reach high levels of efficiency at higher temperatures. I know I am telling Noah about the flood here but I am going somewhere. Mr. Lipman just referred to two problems being financing and licensing. NRC requires approximately 20 years to develop a regulatory pathway for an advanced reactor design like Transatomic's and your company is in that process. Now, according to Businessweek, they had an article on—Bloomberg Businessweek I think it is—on your company. How long have you been doing this? Let me just ask you that question.

Dr. DEWAN. I actually started the company with my cofounder back in 2011 actually.

Mr. WEBER. Okay.

Dr. DEWAN. For the first two years it was when we were in the middle of our Ph.D. program so we have been full-time just for the past year-and-a-half.

Mr. WEBER. Okay. And your cofounder's name?

Dr. DEWAN. Mark Massie.

Mr. WEBER. Mark Massie, okay.

So for you guys has it been a nightmare? I mean there is no clear predictable legal process and permitting process. How does that work for you?

Dr. DEWAN. It has been tricky to say the least. And thank you so much for this question. It is an issue that I spend a great deal of time thinking about.

We believe that ultimately we will be able to find a regulatory pathway for this type of advanced reactor technology in the United States on time scales shorter than the 20 years currently estimated. We feel it is a necessity if the United States wants to take advantage of this molten salt technology that was first developed in this country back in the 1960s, though it is a very tricky path.

Currently, there is no way for us to build a prototype facility or move beyond the laboratory-scale work that we are currently doing. We want more than anything to do this in the United States but we have been forced to keep an open mind with respect to the other pathways we could take.

Mr. WEBER. Canada was mentioned earlier.

Dr. DEWAN. Yes.

Mr. WEBER. According to the Bloomberg Businessweek, you all started in February of 2010. You all decided to fix what is wrong with nuclear reactors.

Dr. DEWAN. Back in spring, summer 2010 was when we first started thinking about—very broadly about advanced reactor designs and how you can do such extraordinary things with all different types of advanced reactors, achieve very high burnouts, produce very little waste.

Mr. WEBER. When did you form your company?

Dr. DEWAN. We incorporated in spring of 2011 actually on the 25th anniversary of Chernobyl.

Mr. WEBER. So 3–1/2 years ago?

Dr. DEWAN. Yes.

Mr. WEBER. So in 3–1/2 years is there anything in that process you would do differently? Can you be very specific about dealing with our agencies?

Dr. DEWAN. So at this point we have been having informal conversations with people at the NRC. We haven't started a—we are not in a position yet to start an application process. We are not in a position yet to even start the pre-application process. That is also the point at which it starts being very, very expensive to engage the NRC once you move beyond informal conversations.

Mr. WEBER. And I don't mean to pry and you may not be at liberty to answer this but have you sought out investors?

Dr. DEWAN. Oh, yes. We have actually raised a round of funding so far from Founders Fund based in San Francisco. They are actually one of the main early investors in SpaceX so they are one of the few VC firms out there that is interested in longer timescale, higher risk, higher reward technology. Otherwise, for the reasons I had mentioned in my testimony, it can be very, very tricky to get private investments in nuclear.

Mr. WEBER. And have they been reluctant because of the permitting and that process?

Dr. DEWAN. A large number of the VC firms that we talked to before we started connecting with Founders Fund, a lot of the other firms were very concerned about the regulatory uncertainty.

And it is not so much the high cost. I feel like if I could tell them—if I could tell potential investors it will cost $200 million just for the regulatory fees in addition to however much it would cost for engineering of the prototype plant, I feel like I could get private investment for that. But when I talk to people and I say, well, it could be 200 million, it could be 100 million, it could be 600 million, I honestly don't know, there are no data points, no one knows, that—

Mr. WEBER. No predictability.

Dr. DEWAN. That isn't something that I can sell to anyone.

Mr. WEBER. We hope to be able to help with that.

And I yield back.

Chairwoman LUMMIS. I thank the gentleman.

And I am going to ask some questions. We have been called to votes but I think that if you will each try to limit your answers to about a minute, you should have an opportunity to respond.

I would like to ask each of you the same question. I am going to start with Dr. Finan and go down the line. And I would like to ask you what can Congress do to assist your efforts to improve licensing processes, to expedite licensing processes, and to help the private sector move forward with potential technologies in this area under discussion today? Dr. Finan?

Dr. FINAN. Thank you.

The NRC operates on a fee-recovery basis. They are required to recover 90 percent of their costs from fees that are paid by operating reactors, and those fees aren't funds that can be used to support regulatory research into an advanced reactor process. So one thing that Congress could do would be to allocate funds to NRC that would be outside of that fee-recovery basis so that they could work on this R&D work and work on developing the groundwork that is needed for innovation and advanced reactor licensing.

Chairwoman LUMMIS. Thank you.

Mr. McGough, same question.

Mr. McGOUGH. Thank you.

So the licensing process that Dr. Finan referred to, we have been involved with the NRC since April of 2008, so we have been paying those bills for a very long time and they are very expensive. To receive our design certification through that point when it will be completed in about 2020, we will have spent $530 million on that process. So it is very expensive.

We need the NRC to issue to us something referred to as a design-specific review standard, which is basically a handshake about how our application will be reviewed. Without that, we are developing an application somewhat blindfolded, without a pre-agreement about when we submit it in this fashion, it will be expeditiously reviewed on the agreed-on 39-month schedule. Even—and that predictability is better than no predictability, as Dr. Dewan referred to. So it is really important to us that we have the NRC dedicated proper resources reviewing those applications in an expeditious fashion.

Chairwoman LUMMIS. Thank you, Mr. McGough.

Dr. Dewan, same question.

Dr. DEWAN. Thank you.

And my answer ties in very closely to what Dr. Finan and Mr. McGough were saying, that what would be most useful would be to encourage the NRC to move to a more staged licensing process similar, as was said before, to a pharmaceutical biotech licensing process where there are multiple stages where designs can get early feedback on the viability of their design through a regulatory process in the United States.

Chairwoman LUMMIS. And Mr. Lipman, finally, same question.

Mr. LIPMAN. Yes. I concur especially with Dr. Finan's suggestion and I might add to it, and that is that Congress could establish a budget line item that is particularly allocated to the review of advanced reactor concepts. There is simply not the mandate for NRC to do that based under the model that Dr. Finan suggested. So that would be helpful very concretely.

Also, the continued investment in many of the programs Dr. Lyons mentioned and some he didn't—or one he didn't—is the LWR Sustainability Program, which allows for advanced work in materials and other aging phenomena that keep our current fleet going.

And lastly, perhaps investment under the laboratory system of advanced materials test reactors. You know, all of these technologies very often depend on behavioral properties of metallurgical phenomena and so reactors at the national laboratories can test these things, their data goes into licensing and can expedite the licensing process. So focus on the fleet and focus on the advanced concepts. Thank you.

Chairwoman LUMMIS. I thank this panel and these witnesses for your valuable testimony and also the Members for their questions.

The Members of the Committee may have additional questions for you and you may receive those questions in writing. We would ask you to respond in writing. The record will remain open for two weeks for additional comments and written questions from Members. Hopefully, you will not get them on the 24th of December so you will be responding to questions on Christmas day.

The witnesses, with our gratitude, are excused.

I would like to ask the staff to prepare a written summary of the last responses that these witnesses gave to that question about what we can do and give it to Chairman Smith so going forward he will know what was recommended for future action or attention by this Committee going forward.

Again, I want to thank our panel and I want to thank you for your wonderful work on our nation's behalf in this important area of research and development.

With that, this hearing is adjourned.

[Whereupon, at 11:53 a.m., the Subcommittee was adjourned.]

Appendix I

Responses by the Hon. Peter Lyons

QUESTIONS FROM CHAIRMAN CYNTHIA LUMMIS

Q1. Dr. Lyons, during the hearing on December 11th, 2014 titled "The Future of Nuclear Energy," you testified that the Office of Nuclear Energy will continue its work related to TRISO fuel. Could you please elaborate on your previous answer and explain how DOE can best make use (of) its investment in TRISO fuel R&D going forward?

A1. Yes. The TRISO coated particle fuel research and development program has successfully completed the initial post irradiation examination of laboratory produced fuel and is moving ahead with the fabrication of TRISO coated particle fuel using prototype commercial scale equipment for the next series of irradiation tests. To date, the experimental work has confirmed TRISO coated particle fuel's ability to retain fission products at temperatures well beyond postulated accident conditions for high temperature gas-cooled reactors. Given the robust nature of the TRISO coated particle fuel, it is also being considered as a fuel form for use in the Fluoride High-temperature Reactors being explored through our Nuclear Energy University Program and is also being examined within the Advanced Fuels program, specifically in the development of accident tolerant fuels for light water reactors.

Q2. Dr. Lyons, you testified that you were not aware of any fundamental impediments for private developers to enter into cooperative R&D agreements with DOE laboratories for the purpose of constructing prototype reactors, but you also noted that there would be a number of challenges related to this concept.

  a. Could you please describe to (the) greatest degree of specificity possible, the challenges you expect for these prospective partnerships?
  b. What additional authority, if any, does DOE require to host a private developer of a prototype reactor at a DOE lab site?

A2 (a).

Historically, the Department of Energy and its predecessors have built over fifty reactors on federal government lands. Challenges facing the construction of future reactors, including those for private developers, may, depending on the facts, include negotiation of cost-share arrangements (e.g., percentage, schedule), establishing advanced reactor licensing criteria for use by the Nuclear Regulatory Commission, compliance with applicable Department of Energy Orders, compatibility of the proposed reactor with the ongoing mission of the site, interface with state/regional/local stakeholders, completion of any required National Environmental Policy Act reviews, protection of intellectual property, integration of needed security and emergency planning into existing site infrastructure, disposal of any generated low level radioactive waste, disposition of used fuel, and long term decommissioning and deconstruction requirements, including equipment disposition.

A2 (b).

At present, we do not believe that additional legislative authority is needed; however, a final determination can be made once specific proposals have been received and evaluated.

Responses by Dr. Ashley Finan

Q1.   Dr. Finan, can you elaborate on what initial steps the Nuclear Regulatory Commission (NRC) should take to develop a workable regulatory framework for non-light water reactor designs?

A1.   Before the NRC can make a commitment to developing a workable regulatory framework for non-light water reactor designs, it will need some resources allocated to that task. Congress can appropriate those funds directly to NRC for that purpose, or to DOE or another third party who could then pay the NRC's fees to work on these topics. Ideally, Congress should appropriate the funds to NRC with direction to use them towards the development of a workable regulatory framework for non-light water reactor designs. This may be best accomplished by adjusting the fee-recovery structure of the NRC so that a larger percentage of the budget can be derived outside user fees, allowing the NRC to work on more forward-looking issues that are important for long term innovation in nuclear energy.

NRC's approach could be based on prior work at the Commission. For example, in 2012, the NRC's Risk Management Task Force published a report, NUREG-2150: "A Proposed Risk Management Regulatory Framework." The Task Force, led by Commissioner George Apostolakis, was chartered "to develop a strategic vision and options for adopting a more comprehensive and holistic risk-informed, performance-based regulatory approach for reactors, materials, waste, fuel cycle, and transportation that would continue to ensure the safe and secure use of nuclear material." Another relevant report of prior work by the NRC is NUREG-1860: "Feasibility Study for a Risk-Informed and Performance-Based Regulatory Structure for Future Plant Licensing" (2007), developed by the Office of Nuclear Regulatory Research. These lines of inquiry would address some of the challenges associated with the prescriptive nature of the NRC's existing guidelines for licensing light water reactors.[1]

Another challenge for advanced reactors is the "all or nothing" structure of the licensing process. That structure imposes a lengthy, costly, and most importantly, unpredictable process for non-light-water reactors. The NRC can take steps to address this by exploring what mechanisms might exist for introducing licensing stages. These stages would provide feedback to developers, investors, and potential

---

[1] In its January, 2012 "Report to the Secretary of Energy," the Blue Ribbon Commission on America's Nuclear Future wrote:
"One area where the Commission recommends increased effort involves ongoing work by the NRC to develop a regulatory framework for advanced nuclear energy systems. Such a framework can help guide the design of new systems and lower barriers to commercial investment by increasing confidence that new systems can be successfully licensed. **Specifically, the Commission recommends that adequate federal funding be provided to the NRC to support a robust effort in this area**. We also support the NRC's risk-informed, performance-based approach to developing regulations for advanced nuclear energy systems...."

users on whether the technology was on the right path to meeting NRC criteria, and thus whether it deserved further investment and development. Receiving that feedback only after a decade and investments on the order of $1 Billion is too high a barrier to innovation. This staged process could involve the use of the existing topical report mechanism in a more formalized and systematic way, or issuing a "statement of licensibility," as is done in some other countries. In developing such a staged pathway, it would be important to collaborate closely with the innovators and investors who would use this process.

Q2.   How do you envision the Department of Energy assisting the NRC in this process?

A2.   The Department of Energy has some experience working with NRC on non-light-water licensing issues, recently through the Next Generation Nuclear Plant project and through the Advanced Reactor Design Criteria initiative. DOE's experience and expertise in advanced reactors generally and advanced reactor licensing in particular make it a useful partner in efforts to develop non-light-water reactor regulations. Potential users and developers of the technologies should also be involved in the process.

DOE could play a key role in working with NRC to clarify and update, as necessary, NRC regulations for non-power reactors, especially if those regulations might be used to oversee reactors on DOE sites. This would be particularly applicable in a situation where DOE was working to enable a "test-bed" facility where private developers could build prototype reactors for new designs.

Responses by Mr. Mike McGough

LO-0115-1079

January 29, 2015

The Honorable Cynthia Lummis, Chair
Subcommittee on Energy
Committee on Science, Space and Technology
U.S. House of Representative
2321 Rayburn House Office Building
Washington, DC 20515-6301

Dear Madame Chair:

RE: Response to testimony during the hearing on December 11th, 2014 titled at "The Future of Nuclear Energy"

Thank you for the opportunity to testify before the Committee on Science, Space and Technology hearing on the "Future of Nuclear Energy" on Thursday, December 11, 2014.

Thank you also for the additional questions, for which I have provided, answers herein, as follows:

   a. Do you believe NRC currently has the staff and resources it needs to review a forthcoming NuScale design application?

On the basis of our extensive pre-application engagements to date, NuScale believes that the NRC has the staff and resources (including the contractors it uses) with the technical capability to review our design certification application and reach a safety conclusion for our design, provided that these resources are assigned to our review. We believe NRC needs to give the NuScale DCA sufficient priority and seek the funding needed (currently estimated at approximately 1% of the NRC budget) to complete its review within 39 months. Further, it would be helpful to the review if the NRC ensured those staff who have been engaging us in pre-application and are familiar with our design were retained for the review of the application upon submittal.

   b. What would it mean to NuScale's prospective customers if the NRC were unable to complete its review within 39 months?

NuScale has extensive interaction with our prospective customer base, which must consider decisions regarding non-carbon emitting baseload power additions many years in advance. Our first project, the Utah Associated Municipal Power Authority Carbon Free Power Project, is targeted for operation in 2024. Since the current forecast for the receipt of the NuScale design certification is the first half of 2020, major financial commitments must be made by UAMPS, and other utilities eager to deploy our SMR technology, long before the completion of the NRC review process. These commitments may coincide with the planned removal from service of other sources of baseload power supply, which cannot be taken out of service without certainty of the replacement source. If the NRC is not able to meet a predictable review schedule it will severely limit the ability of our customer base to select our technology for their needs. Additionally, the ability for our customers to secure the large amounts of financing necessary to build a capital-intensive project like a NuScale facility is contingent upon lenders having mitigated schedule and cost risks of which regulatory risk is a significant component. Importantly, this is not simply a matter of a day-for-day delay—once a customer has selected an alternative generation source, the window of opportunity for NuScale will remain closed with that customer until the next generation decision needs to be made, often 5, 10, or 15 years into the future.

c. How much does NuScale plan to spend to successfully complete NRC's 39-month design review process?

NuScale estimates the total cost to develop our technology and complete all design and engineering work is approximately $1 billion. The cost of the design work necessary to prepare NuScale's submission of our DCA to the NRC in 2016, combined with the cost to complete NRC's proposed 39-month review process will contribute between 50-55% of the total costs.

Thank you again for the opportunity to participate in the important work of this committee. Should you desire further information please do not hesitate to contact me at 410-200-5375 or mmcgough@nuscalepower.com

Sincerely,

Mike McGough
Chief Commercialization Officer

Responses by Dr. Leslie Dewan

Hearing Questions for the Record
following
Testimony before the
U.S. House Committee on Science, Space, and Technology
Energy Subcommittee
The Future of Nuclear Energy

Dr. Leslie Dewan
Co-Founder and CEO
Transatomic Power Corporation

1. Dr. Dewan, can you elaborate on what initial steps the Nuclear Regulatory commission (NRC) should take to develop a workable regulatory framework for non-light water reactor designs?

The NRC's initial steps for developing a workable regulatory framework for non-light water reactor designs should focus on working with the DOE to review the DOE's initial submitted guidelines for advanced reactor licensing, and work rapidly to adapt these guidelines into regulations.

A DOE-NRC Joint Initiative for developing licensing guidelines for advanced reactors is currently underway. The DOE headed Phase 1 of the work, adapting existing General Design Criteria from 10 CFR 50 Appendix A and developing new technology-specific design criteria for advanced reactors. In late December 2014, the DOE submitted their final 99-page report to the NRC. The NRC will head Phase 2 of the work, during which they will review the report over the next 2 to 3 years and use it to develop guidelines for advanced reactor designs. The DOE-led team is not requesting rulemaking to implement the guidance proposed by the initiative.

It is crucial that the NRC ultimately adapt these DOE-submitted guidelines into regulations. It could also be of great benefit to the country for the NRC to devote more resources to their Phase 2 review, allowing the work to be conducted more quickly and with a larger staff.

2. How do you envision the Department of Energy assisting the NRC in this process?

The Department of Energy, through the national laboratory system, has immense scientific and engineering expertise in a wide variety of advanced reactor technologies. Any type of viable regulatory pathway will require collaboration between the DOE and NRC. The initial steps for such a framework are already underway, in the form of the DOE-NRC Joint Initiative for developing a licensing framework for advanced reactors.

This work will likely be an iterative process, and will require continues engagement between the two organizations. To encompass the breadth and depth of current and emerging advanced reactor designs, the DOE should be consulted regularly and frequently to ensure that the licensing criteria are performance-based and technology-agnostic, so that the fundamental safety criteria can be applied to all designs.

Responses by Mr. Daniel Lipman

QUESTIONS FOR THE RECORD

Daniel S. Lipman

Executive Director, Policy Development and Supplier Programs

Nuclear Energy Institute

to the

Committee on Science, Space and Technology

Subcommittee on Energy

U.S. House of Representatives

1. Mr. Lipman, what steps could DOE and the NRC take to remove market barriers for the development of small modular reactors and non-light water reactors in the United States?

Small modular reactors are a strategic priority provided they are cost competitive with the large plants and that time to market is shorter. The nuclear industry will need as many technology options as possible moving forward. Because SMRs allow capacity additions in smaller increments, they may be particularly well-suited to regions of the country with low growth in electricity demand without requiring a company to undertake a 7-8 billion dollar gigawatt-scale project in a single step. To make this possible, DOE and NRC could remove market barriers through the following:

- Ensure that DOE continues funding, and if possible increase, the SMR development program with the industry.
  The NRC should continue to enhance the licensing process for SMRs and establish a licensing process for non-light water reactors that provides a clear pathway for success with a defined schedule and cost. Since nuclear requires higher upfront capital costs, this would create a better environment to attract investors.

2. What other actions could Congress take to enable the development of small modular reactors and non-light water reactors in the United States?

The mostly significant challenges facing both small modular reactors and advance reactors are financing and licensing. In order to enable the US to develop SMRs and non-light water reactors, Congress should ensure the following:

- DOE's R&D budget is fully funded with specific attention to identifying the R&D needs of industry (e.g., to support licensing and fuel qualification) and providing the experimental capabilities to meet these needs.

- The DOE explores ways in which the national laboratories can reduce the time to market for advanced designs and/or serve as a test bed for demonstration of advanced concepts.
- The loan guarantee program is functioning and if necessary restructured to allow for investment in advanced, long-term nuclear technologies
- Ensure DOE continues to fund – or even increase – the SMR development cost share program with industry.

www.ingramcontent.com/pod-product-compliance
Lightning Source LLC
Chambersburg PA
CBHW080828180526
45168CB00006B/2608